1 MONTH OF
FREE
READING

at

www.ForgottenBooks.com

By purchasing this book you are eligible for one month membership to ForgottenBooks.com, giving you unlimited access to our entire collection of over 1,000,000 titles via our web site and mobile apps.

To claim your free month visit:

www.forgottenbooks.com/free906425

ISBN 978-0-266-89794-1
PIBN 10906425

Forgotten Books is a registered trademark of FB &c Ltd.
Copyright © 2018 FB &c Ltd.
FB &c Ltd, Dalton House, 60 Windsor Avenue, London, SW19 2RR.
Company number 08720141. Registered in England and Wales.

For support please visit www.forgottenbooks.com

CANADA
DEPARTMENT OF MINES
Hon. Louis Coderre, Minister; R. W. Brock, Deputy Minister.

GEOLOGICAL SURVEY

MEMOIR 54

No. 2, Biological Series

Annotated List of Flowering Plants and Ferns of Point Pelee, Ont., and Neighbouring Districts

BY

C. K. Dodge

OTTAWA
Government Printing Bureau
1914   No. 1366

CONTENTS

ILLUSTRATIONS

Annotated List of Flowering Plants and Ferns of Point Pelee, Ont., and Neighbouring Districts.

INTRODUCTION.

As early as 1882, a distinguished and enthusiastic ornithologist, Wm. E. Saunders, of London, Ont., visited Point Pelee, and, after some investigation, learned that this place was on a direct line of north and south bird migration. Other ornithologists soon became interested, and an association was formed known as the Great Lakes Ornithological Club, the active members being Mr. Saunders, P. A. Taverner, then of Detroit, Mich., now a member of the Geological Survey, Canada, Bradshae H. Swales, of Grosse Isle, Mich., James S. Wallace, and James Fleming, of Ontario. The locality became so interesting that it was thought best to investigate, as far as possible, the general biota along the line of migration, P. A. Taverner suggesting that the writer undertake the listing of the flowering plants, ferns, and fern allies. The first opportunity came in 1910. After visiting Point Pelee it seemed that a careful study of the vegetation would be of interest and value in two ways: first, it might perhaps be of some service in the study of bird migration; second, it would add much to our knowledge of the distribution of the wild plants of western Ontario and Michigan, in which the writer has been engaged for some years. But it was clear that to serve these two purposes well, not only the plants of Point Pelee should be investigated and listed, but also those of the islands in Lake Erie lying immediately south and extending to the Ohio shore, and the whole of Essex county, as far as possible. The work was cheerfully undertaken, and this paper is the result.

LOCALITIES INVESTIGATED.

Essex county, Ontario, occupies a remarkable and important position, being the most southwesterly county in Ontario. It is bounded on the north by Lake St. Clair, on the west by Detroit river, and on the south by Lake Erie; it extends about forty miles east and west, and thirty miles north and south; its surface is generally level, with a high bank along Lake Erie, and a low bank along Lake St. Clair. Point Pelee is at the southeastern corner of the county and projects about nine miles southward into Lake Erie. Lying in Lake Erie, perhaps less than ten miles to the southwest of this extreme point, is Pelee island, a part of Essex county, by far the largest island in the western part of Lake Erie, and containing about 13,000 acres. At the southern end of the island a narrow point, two and a half miles long, extends south towards the Ohio shore. South and west of this island, and not far away, are numerous smaller islands belonging mostly to the state of Ohio, and extending almost in a direct line close to the Ohio shore. The largest of these islands are: Kelley, Put-in-Bay, Middle Bass, Rattlesnake, and Green islands. Projecting far out from the Ohio mainland in a northerly direction towards the islands named, are two or more peninsulas. From Point Pelee, by way of the islands and these peninsulas, lies the natural path of bird migration.

POINT PELEE.

This locality, as far as was noticed, has been less disturbed by man and retains more of its primitive vegetation than any other equal area on the lake and river shores or anywhere in the county. A careful botanical survey of this place being the principal object, it will receive more particular attention. Point Pelee has been well described, and a map given, by P. A. Taverner and B. H. Swales in their work on "Birds of Point Pelee",[1] which appeared in 1908, and no further description will be attempted here than is necessary for botanical discussion. It is a triangular piece of land with its acute angle running

---

[1] Reprinted from the Wilson Bulletin, No. 59, June, 1907.

about nine miles south into Lake Erie, the base at the north being about six miles wide. About nine-tenths of this tract was, and still is, mostly a very wet marsh lying between the

Fig. 1. Index of Point Pelee and vicinity.

east and west beaches. Formerly it was famous for water birds and one of the best spots known for duck shooting; but in recent years, particularly near the base, it has been quite extensively ditched and pumping stations established, draining and fitting a part of it for cultivated crops. Ultimately, most of the marsh will, no doubt, be reclaimed, and already its importance for duck hunting has greatly diminished. With these changed conditions much of the primitive vegetation has disappeared and species adapted to drier ground have taken possession, but here and there many large spots are still to be found showing conclusively the original plant life, which was quite uniform and limited in species.

The wetter part was covered more or less with common cat-tail flag, Indian rice (*Zizania palustris* L.[1]) in great abundance, one of our largest and most striking native grasses, often over ten feet high, swamp horsetail, broad-fruited bur reed, broad-leaved arrow head, rice cut grass, common reed, western bulrush, slender sedge, swamp loosestrife, mermaid weed, rush aster, and tall tickseed sunflower (*Bidens trichosperma* (Michx.) Britton., var. *tenuiloba* (Gray) Britton), and in spots, reed canary grass and northern manna grass (*Glyceria septentrionalis* Hitchc.) are abundant. Within the marsh limits are several ponds and small lakes with characteristic plants, these being several species of waterweeds, wild celery, yellow water lily, white water lily, water shield, and slender najas. On the marsh border, especially along the west side, are, in great abundance, sweet flag, poison sumach, and button-bush, twenty feet high and more. On the east side is a narrow low sandy beach, fringed with shrubs and trees towards the south. On the west side the beach is wider, backed by sand dunes covered with shrubs and trees. It is at first, beginning at the north, a very narrow strip popularly known as "the narrows", growing wider towards the south, until the east and west beaches with their wooded strips come together about two miles north of the extreme southern point, forming from there on to the south, flat sandy ground, covered with timber, but bordered on the east side by a strip of damp rich woods. Outside of the marsh the land is mostly very sandy, though here and there it is under cultivation. The road from Leamington passes along the so-called narrows and

---

[1] Two species of Indian rice have been lately recognized. *Zizania aquatica* L., the smaller form, was not noticed.

terminates at the extreme southern point. Going along this road the last of May, 1910, the writer never witnessed a more beautiful display of the wild columbine in full bloom. On this narrow strip between the west beach and the marsh, the black walnut, one of our finest trees, usually found in rich open woods or on river and creek bottoms, is .frequent and apparently thrifty in pure sand, and the hackberry is very common. These two trees are very plentiful about half-way to the south point. Here also the blue ash was noticed, with its peculiar square stems, and the honey locust mostly on the middle and upper beach, perhaps fifty trees or more quite out of their natural habitat, yet apparently thrifty. The dominant tree of the timbered portion, however, is the red cedar, and it is more or less abundant along the whole west side. About halfway to the extreme southern point is a clump of fair-sized white pine trees, and this pine is occasionally seen in other places. Red oak and yellow-barked oak (*Quercus velutina* Lam.) are common; the chestnut oak occasional.

Prominent among the beach herbaceous plants is the clammy-weed which has crept into gardens and fields and become a pernicious weed. American searocket and sand grass are pretty evenly distributed along both beaches. On the west side, the shrubs, fragrant sumach, and low Juniper, are very abundant and efficacious in holding down the sand dunes and beach sand against the action of the wind. The sea sand reed, one of the best known sand binders, very efficacious in resisting the action of both wind and wave, appears in spots on the east side, but is not anywhere abundant. The long-leaved reed grasses, another noted sand binder, are not at all common. In the strip of rich damp woods on the east side and south of the big marsh, are found the usual trees and herbaceous plants that occur where mesophytic conditions prevail—white ash, black ash, American elm, red maple, basswood, cottonwood, bur oak, swamp white oak, sycamore, peach-leaved willow, Virginian knotweed, bloodroot, purple cress, and cardinal flower. The honey locust, shrubby trefoil, locally called wahoo, western prickly pear, rough-leaved cornel, red mulberry, and papaw, very probably reach their northern limit here, the last two far disappearing. It is perhaps quite remarkable that only one member of the Heath family, bearberry, was noticed; it is quite plentiful and serves as a very good sand binder, helping to hold the sand dunes in place on the west side. The American cranberry existed in the big marsh before drainage and destruction by fire.

Pelee island, by far the largest island in the western part of Lake Erie, was looked over as carefully as time would permit. It is generally flat, the rock often having only a thin covering of soil. Much of the land has been under cultivation. The soil is generally good, and formerly the grape and peach were extensively cultivated; but both of these have in recent years been neglected, and much attention paid to raising tobacco. Here the Kentucky coffee-tree, redbud and trumpet creeper, the last quite common, seem to reach their northern limit.

A knowledge of the plants on the other islands has been made possible by the careful botanical examination of all those islands south of the International Boundary to the Ohio shore and the points and peninsulas of the shore, by Prof. E. L. Moseley, of Sandusky, Ohio, who published his conclusions and catalogue of plants in 1899.[1] This publication is a work of great value, and the author has kindly permitted me to use the results of his labour. Every plant mentioned as on or about "the islands" or any one of them, along the Ohio shore, is taken from his list.

NUMBER OF SPECIES NOTED IN DIFFERENT LOCALITIES.

The number of plants noted on Point Pelee is 583, on Pelee island 408, on both 623, there being about 40 species on the island not yet found on the point. These figures are only close approximations. On the islands south of the International Boundary, Professor Moseley found 612 species, 176 of which have not yet been found on Point Pelee or the island. The apparent absence of this rather large number of species is owing,

---

[1] A Catalogue of Flowering Plants and Ferns Growing Without Cultivation in Erie county, Ohio, and the peninsula and islands of Ottawa county. By E. L. Moseley.

in the main, to the restricted area examined, for many of th. plants noticed on the islands by Professor Moseley and not found on Point Pelee and the adjacent large island, are known to be frequent and often abundant in western Ontario, such as *Cystopteris bulbifera* (L.) Bernh., *Allium tricoccum* Ait., *Erythronium americanum* Kerr., *Trillium erectum* L., *Ranunculus fascicularis* Muhl., *Hydrastis canadensis* L., *Jeffersonia diphylla* (L.) Pers., *Arabis canadensis* L., and many others.[1]

Professor Moseley detected on Kelley island, alone, 461 species, and on Put-in-Bay island, 439 species, while, as before stated, only 408 were noted and reported on Pelee island, whose area is far greater than the combined area of all the other mentioned islands. It is very evident, therefore, that many of the plants of Pelee island have not yet been noted and reported, showing clearly that there is still work for a local botanist. There should be found on this island about 650 or 700 species growing without cultivation, that is from 242 to 292, or about 300 more than is reported in this paper. It is very probable, however, that no plants will hereafter be found on Pelee island not already reported from Point Pelee or the other islands of Lake Erie to the south, or in other parts of Essex county.

On the preparation of the following list the writer has made use of every available source of information. The nomenclature of "Gray's New Manual of Botany, Illustrated" has been followed, unless otherwise mentioned. Many common names have been taken from Britton and Brown's illustrated work. The writer is much indebted to Agnes Chase, scientific assistant in systematic agrostology, Bureau of Plant Industry, United States Department of Agriculture, for examining the various species of grasses; and to Kenneth K. Mackenzie, of New York city, for inspecting all species of *Cyperaceae*, *Junci*, and many other plants.

---

[1] Many of the above-named plants have since been noticed in other parts of Essex county and inserted in the list.

ANNOTATED LIST.

POLYPODIACEAE (Fern Family.)

Polypodium vulgare L. (Common Polypody.)
Kelly island. Scarce.

Phegopteris hexagonoptera (Michx.) Fée. (Broad Beach Fern.)
In rich woods near Windsor. (Burgess.)

Phegopteris Dryopteris (L.) Fée. (Oak Fern.)
Frequent about Windsor. (F. P. Cravin.)

Adiantum pedatum L. (Maidenhair.)
Common in rich shaded ground about Windsor. Ohio shore.

Pteris aquilina L. (Common Brake.)
Frequent on Point Pelee, Pelee island, and in Essex county generally.

Pellaea atropurpurea (L.) Link. (Cliff Brake.)
Kelley and Put-in-Bay islands, and Ohio shore.

Asplenium Trichomanes L. (Maidenhair Spleenwort.)
Ohio shore.

Asplenium platyneuron (L.) Oakes. (Ebony Spleenwort.)
Ohio shore.

Asplenium angustifolium Michx. (Narrow-leaved Spleenwort.)
Rich woods near Amherstburg. (Maclagan.) Ohio shore.

Asplenium Filix-femina (L.) Bernh. (Lady Fern.)
Ohio shore and no doubt throughout Essex county.

Camptosorus rhizophyllus (L.) Link. (Walking Fern.)
Kelley island and Ohio shore.

Polystichum acrostichoides (Michx.) Schott. (Christmas Fern.)
Frequent about Windsor. (F. P. Cravin.)

Aspidium Thelypteris (L.) Sw. (Marsh Shield Fern.)
Very common in marshy open ground or slightly shaded places at Pelee islands and Ohio shore.

Aspidium noveboracense (L.) Sw. (NEW YORK FERN.)
Low woods and thickets near Windsor.

Aspidium marginale (L.) Sw.
Frequent in rich woods about Windsor. (F. P. Cravin.)

Aspidium Goldianum Hook. (GOLDIE'S FERN.)
In rich woods near Amherstburg. (Maclagan.)

Aspidium cristatum (L.) Sw. (CRESTED SHIELD FERN.)
Occasional in swampy places on the east side of Point
Pelee and about Windsor. No doubt frequent through-
out other parts of Essex county.

Aspidium spinulosum (O. F. Müller) Sw. (SPINULOSE SHIELD
FERN.)
Frequent at Point Pelee in rich woods, islands, and Ohio
shore. No doubt frequent throughout Essex county.

Aspidium spinulosum (O. F. Müller) Sw., var. **intermedium**
(Muhl.) D. C. Eaton. (SPINULOSE SHIELD FERN.)
In rich woods on the east side of Point Pelee and on Pelee
island. Probably throughout Essex county.

Cystopteris bulbifera (L.) Bernh. (BULBLET CYSTOPTERIS.)
Lake Erie islands.

Cystopteris fragilis (L.) Bernh. (BRITTLE FERN.)
Kelley island and Ohio shore.

Onoclea sensibilis L. (SENSITIVE FERN.)
Common in damp open ground, woods, and thickets on
the east side of Point Pelee and on Pelee island.
Not noticed on the other islands, but common on the
Ohio shore.

Onoclea Struthiopteris (L.) Hoffm. (OSTRICH FERN.)
Frequent in Essex county. (F. P. Cravin.)

OSMUNDACEAE (FLOWERING FERN FAMILY.)

Osmunda regalis L. (FLOWERING FERN.)
Common in damp open ground about Windsor.

Osmunda Claytoniana L. (CLAYTON FERN.)
Frequent about Windsor.

Osmunda cinnamomea L. (CINNAMON FERN.)
Common about Windsor.

OPHIOGLOSSACEAE (Adder's Tongue Family.)

Ophioglossum vulgatum L. (Adder's Tongue.)
Cedar point, Ohio shore. Probably frequent in Essex county, but overlooked.

Botrychium simplex E. Hitchcock. (Little Grape Fern.)
Noticed at Cedar point, Ohio shore.

Botrychium ramosum (Roth) Aschers. (Matricarv Grape Fern.)
At Cedar point, Ohio shore.

Botrychium virginianum (L.) Sw. (Rattlesnake Fern.)
Frequent in rich woods and thickets on the east side of Point Pelee and on Pelee island. Very probably common throughout Essex county.

EQUISETUM (Horsetail Family.)

Equisetum arvense L. (Common Horsetail.)
In dry or damp open ground at Point Pelee and on Pelee island. Common throughout Essex county. Kelley island and Ohio shore.

Equisetum sylvaticum L. (Wood Horsetail.)
Occasional in damp shaded places on the east side of Point Pelee and on Pelee island. No doubt throughout Essex county.

Equisetum fluviatile L. (Pipes.) (Swamp Horsetail.)
Common about ponds, in ditches, shallow water, and wet places, especially on the big marsh at Point Pelee. Also on Pelee island, and probably throughout Essex county.

Equisetum laevigatum A. Br. (Smooth Scouring Rush.)
Reported as noticed along road-sides near Windsor. According to the late A. A. Eaton, this species is very doubtful in western Ontario and Michigan, it being often mistaken for E. hyemale intermedium. A. A. Eaton.

Equisetum hyemale L. (Scouring Rush.)
> Frequent in dry open places among cedars at Point Pelee and on Pelee island. No doubt frequent throughout Essex county.

Equisetum hyemale L., var. **intermedium** A. A. Eaton. (Scouring Rush.)
> Noticed near Windsor. According to the late A. A. Eaton, this is often confused with E. Laevigatum.

Equisetum hyemale L., var. **robustum** (A. Br.) A. A. Eaton. (Stout Scouring Rush.)
> Put-in-Bay and Kelley islands and Ohio shore.

Equisetum variegatum Schleich. (Variegated Equisetum.)
> Abundant on borders of ponds, east side of Point Pelee. Also islands and Ohio shore.

SELAGINELLACEAE (Selaginella Family.)

Selaginella apus (L.) Spring. (Creeping Selaginella.)
> Damp open places along Detroit river. Probably throughout Essex county, but overlooked.

TAXACEAE (Yew Family.)

Taxus canadensis Marsh. (American Yew.)
> Among cedars and pines on the west side of Point Pelee. Apparently rare. (Wallace Tilden.) Rocky shores of the islands.

PINACEAE (Pine Family.)

Pinus Strobus L. (White Pine.)
> Usually scattering at Point Pelee. A grove of large trees about halfway down to extreme point. Occasional in other parts of the county. Ohio shore.

Pinus sylvestris L. (Scotch Fir.)
> Fine specimens in cultivation on north shore of Lake Erie, but not noticed as spreading.

Larix laricina (Du Roi) Koch. (AMERICAN LARCH.)

>Occasional in swampy places, but formerly more abundant.

Juniperus communis L. (COMMON JUNIPER.)

>A few trees at Point Pelee might perhaps be taken for the species. Ohio shore.

Juniperus communis L., var. **depressa** Pursh. (LOW JUNIPER.)

>Abundant on the west side of Point Pelee along the beach and on near-by sand ridges, acting as an efficient sand binder against the action of the wind. Fruit usually very abundant. Also on the extreme southern point of Pelee island. Ohio shore.

Juniperus virginiana L. (RED CEDAR.)

>Abundant and the dominant tree in many places on the west side of Point Pelee from "the narrows" to the extreme southern point. Also on Pelee island and scattering along north shore of Lake Erie to the Detroit river. Abundant on the islands and Ohio shore.

TYPHACEAE (CAT-TAIL FAMILY.)

Typha latifolia L. (COMMON CAT-TAIL.)

>Abundant in ditches and the big marsh at Point Pelee and on Pelee island, and in swampy places throughout Essex county. Also islands and Ohio shore.

Typha angustifolia L. (NARROW-LEAVED CAT-TAIL.)

>Frequent and often plentiful on borders of the big marsh at Point Pelee. North Bass island and Ohio shore.

SPARGANIACEAE (BUR REED FAMILY.)

Sparganium eurycarpum Engelm. (BROAD-FRUITED BUR REED.)

>Very common in and about the big marsh, in ditches and very wet places at Point Pelee and on Pelee island. Middle Bass island and Ohio shore.

Sparganium americanum Nutt., var. **androcladum** (Engelm) Fernald and Eames. (BRANCHING BUR REED.)

Occasional in and about the big marsh at Point Pelee. Middle Bass island and Ohio shore.

NAJADACEAE (PONDWEED FAMILY.)

Potamogeton natans L. (COMMON FLOATING PONDWEED.)

Ponds and small lakes in the big marsh at Point Pelee an : on Pelee island. Common about the other islands.

Potamogeton americanus C. and S. (LONG-LEAVED POND-WEED.)

Occasional in water on the east side of Point Pelee. Common about the islands.

Potamogeton amplifolius Tuckerm. (LARGE-LEAVED POND-WEED.)

In water along Ohio shore.

Potamogeton heterophyllus Schreb. (VARIOUS-LEAVED PONDWEED.)

Occasional in water on east side of Point Pelee. Near Windsor. (Macoun.) Ohio shore.

Potamogeton heterophyllus Schreb., forma **maximus** Morong. (VARIOUS-LEAVED PONDWEED.)

North Bass island and Ohio shore.

Potamogeton lucens L. (SHINING PONDWEED.)

In water about the islands and along Ohio shore.

Potamogeton Richardsonii (Benn.) Rydb. (RICHARDSON PONDWEED.)

In water at the north end of east side of Point Pelee. Islands and Ohio shore.

Potamogeton perfoliatus L. (CLASPING-LEAVED PONDWEED.)

In water along Ohio shore.

Potamogeton zosterifolius Schumacher. (EEL-GRASS POND-WEED.)

Occasional in the big ditches at north end of Point Pelee on east side. Common about the islands and along Ohio shore.

Potamogeton Hillii Morong.　(HILL's PONDWEED.)
> Along the Ohio shore.

Potamogeton Friesii Rupr.　(FRIES' PONDWEED.)
> In big ditches on east side of north end of Point Pelee.
> Also about Put-in-Bay island and along Ohio shore.

Potamogeton pusillus L.　(SMALL PONDWEED.)
> In water along Ohio shore.

Potamogeton foliosus Raf.　(LEAFY PONDWEED.)
> Frequent at the north end of Point Pelee on the east
> side. Also about Put-in-Bay and North Bass island.

Potamogeton foliosus Raf., var. **niagarensis** (Tuckerm.)
Morong.　(LEAFY PONDWEED.)
> About North Bass island and along Ohio shore.

Potamogeton pectinatus L.　(FENNEL-LEAVED PONDWEED.)
> Occasional at north end of Point Pelee on the east side.
> Abundant about the islands and along Ohio shore.

Potamogeton interruptus Kitaibel.　(INTERRUPTED POND-
WEED.)
> About Put-in-Bay island.

Najas flexilis (Willd.) Rostk. and Schmidt.　(SLENDER NAJAS.)
> In big ditches at north end of Point Pelee on east side.
> Also about the islands.

Najas flexilis (Willd.) Rostk. and Schmidt, var. **robusta**
Morong.　(LARGER NAJAS.)
> In water along Ohio shore.

Najas gracillima (A. Br.) Magnus.　(THREAD-LIKE NAJAS.)
> In water along Ohio shore.

JUNCAGINACEAE (ARROW GRASS FAMILY.)

Triglochin maritima L.　(SEASIDE ARROW GRASS.)
> Frequent at Point Pelee in wet sand or marshy open
> ground in and on borders of big marsh.

ALISMACEAE (WATER-PLANTAIN FAMILY.)

Sagittaria latifolia Willd.　(BROAD-LEAVED ARROW-HEAD.)
(SWAN-ROOT.)
> Common at Point Pelee in ditches and wet places on the
> big marsh. Islands and Ohio shore.

Sagittaria arifolia Nutt. ((ARUM-LEAVED ARROW-HEAD.)
Along the Ohio shore.

Sagittaria heterophylla Pursh. (SESSILE-FRUITING ARROW-HEAD.)
Borders of small lakes in big marsh. Put-in-Bay island and Ohio shore.

Sagittaria graminea Michx. (GRASS-LEAVED SAGITTARIA.)
Along Ohio shore.

Alisma Plantago-aquatica L. (WATER PLANTAIN.)
Abundant in ditches and very wet places at Point Pelee and throughout Essex county. Also islands and Ohio shore.

HYDROCHARITACEAE (FROG'S BIT FAMILY.)

Elodea canadensis Michx. (WATER-WEED.)
Common in the ponds and small lakes of the big marsh at Point Pelee and on Pelee island. Also Kelley and Put-in-Bay islands and Ohio shore.

Vallisneria spiralis L. (WILD CELERY.)
Common at Point Pelee in the ponds and small lakes of the big marsh. Abundant about the islands and along Ohio shore.

GRAMINEAE (GRASS FAMILY.)

Andropogon scoparius Michx. (BROOM BEARD GRASS.)
Occasional at Point Pelee in open dry ground, and on Pelee island. Ohio shore.

Andropogon furcatus Muhl. (FORKED BEARD GRASS.)
Occasional at Point Pelee in dry open ground. Ohio shore.

Sorghastrum nutans (L.) Nash. (INDIAN GRASS.)
On sandy open ground at Point Pelee. Apparently rare. Ohio shore.

Digitaria humifusa Pers. (SMALL CRAB GRASS.)
Occasional at Point Pelee in sandy ground and on Pelee island. North Bass island and Ohio shore.

Digitaria sanguinalis (L.) Scop. (CRAB GRASS.)

A common weed at Point Pelee and on Pelee island. The other islands and Ohio shore.

Panicum capillare L. (OLD-WITCH GRASS.)

A common weed in gardens, fields, and waste places at Point Pelee and on Pelee island. The other islands and Ohio shore.

Panicum flexile (Gattinger) Scribn. (WIRY PANICUM.)

Common in moist sandy ground at the north end of Point Pelee on east side.

Panicum philadelphicum Bernh. (WOOD WITCH GRASS.)

Noticed at Point Pelee in dry sandy ground among red cedars. (A. B. Klugh.) Ohio shore.

Panicum miliaceum L. (EUROPEAN MILLET.)

Ohio shore. Probably also as an escape in Essex county.

Panicum virgatum L. (SWITCH GRASS.)

Usually in tufts on dry sandy ground. Abundant along the upper east beach of Point Pelee. Also Kelley island and Ohio shore.

Panicum agrostoides Spreng. (AGROSTIS-LIKE PANICUM.)

North Bass island.

Panicum depauperatum Muhl. (STARVED PANICUM.)

Occasional at Point Pelee in dry poor ground. Ohio shore.

Panicum dichotomum L. (FORKED PANICUM.)

Frequent in woods at Point Pelee. (Burgess.) Also on islands.

Panicum huachucae Ashe. (HAIRY PANICUM.)

Occasional in sandy ground at Point Pelee. No doubt to be found throughout Essex county.

Panicum scoparium Lam. (SCRIBNER'S PANICUM.)

Occasional at Point Pelee in dry open ground. Probably throughout Essex county. On islands and Ohio shore.

Panicum clandestinum L. (HISPID PANICUM.)

Cedar point, Ohio shore.

Echinochloa crusgalli (L.) Beauv. (BARNYARD GRASS.)

A weed at Point Pelee and probably throughout Essex county, in gardens, fields, and waste places, preferring damp ground. Also on islands and Ohio shore.

Echinochloa Walteri (Pursh) Nash. (SALT-MARSH COCK-SPUR GRASS.)

Common in damp places near creeks flowing into Lake Erie along north shore. Ohio shore.

Setaria glauca (L.) Beauv. (FOXTAIL.)

A common weed throughout Essex county in gardens and fields. Islands and Ohio shore.

Setaria viridis (L.) Beauv. (GREEN FOXTAIL.)

Throughout Essex county as a weed in cultivated fields. Ohio shore.

Setaria italica (L.) Beauv. (ITALIAN MILLET.)

Occasional as an escape throughout Essex county. On Middle Bass and North Bass islands and Ohio shore.

Cenchrus carolinianus Walt. (SANDBUR.)

Occasional as a roadside weed throughout Essex county. Islands and Ohio shore.

Zizania palustris L. (INDIAN RICE.)

Very abundant and rank in spots on the big marsh at Point Pelee. Also along low wet banks of streams flowing into Lake Erie and about Lake St. Clair. Also on Pelee island, Middle Bass island, and Ohio shore. The smaller form Z. aquatica L. not noticed.

Leersia virginica Willd. (WHITE GRASS.)

Frequent at Point Pelee in rich open woods on the east side. Also Kelley island and Ohio shore.

Leersia oryzoides (L.) Sw. (RICE CUT GRASS.)

Common at Point Pelee along ditches and in wet places, especially in the big marsh. Also on islands and Ohio shore.

Phalaris arundinacea L. (REED CANARY GRASS.)

Occasional at Point Pelee in wet open places, especially in the big marsh. Middle Bass island and Ohio shore.

Hierochloe odorata (L.) Wahlenb. (VANILLA GRASS.)

Grassy places along Detroit river.

Milium effusum L. (MILLET GRASS.)

In rich open woods on the east side of Point Pelee.

Oryzopsis pungens (Torr.) Hitchc. (SLENDER MOUNTAIN RICE.)

In dry open ground about Windsor.

Oryzopsis asperifolia Michx. (WHITE-GRAINED MOUNTAIN RICE.)

In shaded ground near Windsor.

Oryzopsis racemosa (Sm.) Ricker. (BLACK-FRUITED MOUNTAIN RICE.)

Put-in-Bay island. Apparently rare.

Stipa spartea Trin. (PORCUPINE GRASS.)

Sandy ground, Ohio shore.

Muhlenbergia sobolifera (Muhl.) Trin. (ROCK MUHLENBERGIA.)

Ohio shore.

Muhlenbergia sylvatica Torr. (WOOD MUHLENBERGIA.)

Occasional on borders of woods, east side of Point Pelee. Middle Bass island.

Muhlenbergia foliosa Trin. (MINNESOTA MUHLENBERGIA.)

Along edge of big marsh at Point Pelee. Apparently not common.

Muhlenbergia mexicana (L.) Trin. (MEADOW MUHLENBERGIA.)

Edge of woods on east side of Point Pelee. Apparently infrequent. Islands and Ohio shore.

Muhlenbergia racemosa (Michx.) BSP. (MARSH MUHLENBERGIA.)

Frequent at Point Pelee in damp open ground.

Muhlenbergia Schreberi J. F. Gmel. (NIMBLE WILL.)

Common at Point Pelee in dry sandy ground under red cedars and other trees. Islands and Ohio shore.

Brachyelytrum erectum (Schreb.) Beauv. (BRACHYELYTRUM.)

Ohio shore.

Phleum pratense L. (TIMOTHY.)

Occasional everywhere as an escape. Islands and Ohio shore.

Alopecurus geniculatus L., var. **aristulatus** Torr. (FLOAT-ING FOXTAIL.)

In very wet open places at Point Pelee and on Pelee island. Islands and Ohio shore.

Sporobolus vaginiflorus (Torr.) Wood. (SHEATHED RUSH GRASS.)

Kelley and Put-in-Bay islands and Ohio shore.

Sporobolus neglectus Nash. (SMALL RUSH GRASS.)

Ohio shore.

Sporobolus cryptandrus (Torr.) Gray. (SAND DROPSEED.)

Frequent at Point Pelee on the beaches and sand ridges. Ohio shore.

Agrostis alba L. (RED TOP.)

Common at Point Pelee in damp open meadow-like ground and occasional in dry sandy places. Common on the islands and Ohio shore.

Agrostis hyemalis (Walt.) BSP. (HAIR GRASS.)

Occasional at Point Pelee in dry open ground. Put-in-Bay and Middle Bass islands and Ohio shore.

Agrostis perennans (Walt.) Tuckerm. (THIN GRASS.)

Ohio shore.

Calamovilfa longifolia (Hook.) Hack. (LONG-LEAVED REED GRASS.)

On the upper beach at Point Pelee but not abundant. When abundant it is one of the best sand binders against the action of wind.

Calamagrostis canadensis (Michx.) Beauv. (BLUE-JOINT GRASS.)

Abundant at Point Pelee in spots in the big marsh and on Pelee island. Middle Bass and North Bass islands and Ohio shore.

Ammophila arenaria (L.) Link. (SEA SAND-REED.)

A characteristic beach grass. Noticed as plentiful in spots on the east side of Point Pelee, but infrequent on the west side. When abundant, it is one of the most efficient sand binders against both wind and wave. Ohio shore.

Cinna arundinacea L. (WOOD REED GRASS.)

In damp woods on the east side of Point Pelee. Apparently infrequent. Ohio shore.

Sphenopholis obtusata (Michx.) Scribn. (BLUNT-SCALED EATONIA.)

North Bass island and Ohio shore.

Sphenopholis pallens (SPRENG.) Scribn. (PENNSYLVANIA EATONIA.)

Islands and Ohio shore.

Koeleria cristata (L.) Pers. (KOELERIA.)

Occasional at Point Pelee on dry sandy ground. Ohio shore.

Danthonia spicata (L.) Beauv. (COMMON WILD-OAT GRASS.)

Frequent at Point Pelee on dry sandy ground. Put-in-Bay island and Ohio shore.

Spartina Michauxiana Hitchc. (SLOUGH GRASS.)

Occasional at Point Pelee in marshy open ground and on Pelee island, Middle Bass island, and Ohio shore. Formerly included with *S. cynosuroides* (L.) Roth. which is now claimed to be an eastern species.

Bouteloua curtipendula (Michx.) Torr. (RACEMED BOUTELOUS.)

Ohio shore.

Eleusine indica Gaertn. (GOOSE GRASS.)

Ohio shore.

Phragmites communis Trin. (REED.)

Abundant in spots on the big marsh at Point Pelee, and on Pelee island. Ohio shore.

Tridens flavus (L.) Hitchc. (TALL RED TOP.)

Ohio shore.

Triplasis purpurea (Walt.) Chapm. (SAND GRASS.)

Common in sand on the beach at Point Pelee, Kelley island, and Ohio shore.

Eragrostis hypnoides (Lam.) BSP. (CREEPING ERAGROSTIS.)

Ohio shore.

Eragrostis Frankii (Fisch, Mey., and Lall.) Steud. (FRANK'S ERAGROSTIS.)

Ohio shore.

Eragrostis capillaris (L.) Nees. (CAPILLARY ERAGROSTIS.)
Ohio shore.

Eragrostis pilosa (L.) Beauv. (TUFTED ERAGROSTIS.)
Plentiful at Point Pelee in dry sandy ground. Often abundant on railway embankments and along highways. Common throughout Essex county. Includes *E. purshii* Schrad. Kelley island.

Eragrostis megastachya (Koeler) Link. (STRONG-SCENTED ERAGROSTIS.)
A weed at Point Pelee in gardens and cultivated fields. Islands.

Eragrostis pectinacea (Michx.) Steud. (PURPLE ERAGROSTIS.)
Cedar point, Ohio shore.

Dactylis glomerata L. (ORCHARD GRASS.)
Occasional at Point Pelee along roads and about dwellings. Pelee island and Ohio shore.

Poa annua L. (LOW SPEAR GRASS.)
About dwellings, especially in lawns, at Point Pelee and on Pelee island. Probably common throughout Essex county.

Poa compressa L. (CANADA BLUE GRASS.)
Common at Point Pelee in dry sandy ground. Islands and Ohio shore.

Poa pratensis L. (JUNE GRASS.)
Common at Point Pelee and on Pelee island. The common grass of open dryish woods, fields, pastures, and lawns. Abundant on the other islands and Ohio shore.

Poa debilis Torr. (WEAK SPEAR GRASS.)
Ohio shore.

Glyceria nervata (Willd.) Trin. (FOWL MEADOW GRASS.)
Common at Point Pelee in damp open places and rich open woods. Islands and Ohio shore.

Glyceria pallida (Torr.) Trin. (PALE MANNA GRASS.)
Ohio shore.

Glyceria septentrionalis Hitchc. (NORTHERN MANNA GRASS.)

Frequent at Point Pelee in ditches, shallow water, and about ponds, especially on the east side of the big marsh. Islands. Formerly included with *G. fluitans* (L.) R. Br.

Festuca octoflora Walt. (SLENDER FESCUE GRASS.)

Occasional at Point Pelee on dry sandy ground. Ohio shore.

Festuca ovina L. (SHEEP'S FESCUE.)

Frequent at Point Pelee in dry sandy ground.

Festuca elatior L. (MEADOW FESCUE.)

Occasional at Point Pelee in meadow-like ground. Islands and Ohio shore.

Festuca nutans Spreng. (NODDING FESCUE.)

Occasional at Point Pelee about farm buildings. Islands of Lake Erie, except Kelley island.

Bromus secalinus L. (COMMON CHESS.)

Occasional at Point Pelee about dwellings and in cultivated grounds. Islands and Ohio shore.

Bromus commutatus Schrad. (VARIABLE BROME GRASS.)

Noticed only in open ground on the west side of Pelee island.

Bromus racemosus L. (UPRIGHT CHESS.)

Roadsides and railway banks at Essex Centre. (Burgess.) Common on the islands.

Bromus ciliatus L. (FRINGED BROME GRASS.)

Occasional at Point Pelee in rich open woods and thickets. Kelley and Rattlesnake islands and Ohio shore.

Lolium perenne L. (COMMON DARNEL.)

Kelley and Put-in-Bay islands.

Agropyron repens (L.) Beauv. (COUCH GRASS.)

Occasional at Point Pelee about farm buildings and in old fields. Probably frequent throughout Essex county. Kelley island and Ohio shore.

Agropyron dasystachyum (Hook.) Scribn. (NORTHERN WHEAT GRASS.)

In sandy ground near shore on Pelee island, but nowhere abundant.

Hordeum jubatum L. (SQUIRREL-TAIL GRASS.)

Occasional at Point Pelee about dwellings. Kelley island and Ohio shore. Becoming common in many parts of western Ontario.

Elymus virginicus L. (VIRGINIA WILD RYE.)

Occasional at Point Pelee on borders of damp woods. Islands and Ohio shore.

Elymus canadensis L. (NODDING WILD RYE.)

Frequent at Point Pelee and Pelee island on the upper beach, but nowhere abundant. The other islands and Ohio shore.

Elymus canadensis L., var. **glaucifolius** (Muhl.) Gray. (GLAUCUS-LEAVED WILD RYE.)

Islands and Ohio shore.

Elymus striatus Willd. (SLENDER WILD RYE.)

Occasional in dryish woods on Pelee island. Kelley island and Ohio shore.

Hystrix patula Moench. (BOTTLE-BRUSH GRASS.)

In rich open woods at Point Pelee. Islands and Ohio shore.

CYPERACEAE (SEDGE FAMILY.)

Cyperus diandrus Torr. (LOW CYPERUS.)

Frequent at Point Pelee in damp open ground, especially about the big marsh. Islands and Ohio shore.

Cyperus rivularis Kunth. (SHINING CYPERUS.)

Ohio shore.

Cyperus Schweinitzii Torr. (SCHWEINITZ'S CYPERUS.)

Frequent at Point Pelee in dry open sandy ground. Common on Ohio shore.

Cyperus esculentus L. (YELLOW NUT GRASS.)

On low ground at Point Pelee. (Burgess.)

Cyperus ferax Rich. (MICHAUX'S CYPERUS.)

In damp sand near Lake Erie shore west of Kingsville. Ohio shore.

Cyperus strigosus L. (STRAW-COLOURED CYPERUS.)

Frequent at Point Pelee in damp, grassy, meadow-like ground. Islands and Ohio shore.

Cyperus Houghtonii Torr. (HOUGHTON'S CYPERUS.)
Frequent along the sandy shore of Lake St. Clair.

Cyperus filiculmis Vahl. (SLENDER CYPERUS.)
Occasional at Point Pelee in very dry open ground. Ohio shore.

Dulichium arundinaceum (L.) Britton. (DULICHIUM.)
Occasional at Point Pelee in wet open places about the big marsh. Ohio shore.

Eleocharis obtusa (Willd.) Schultes. (BLUNT SPIKE RUSH.)
Occasional at Point Pelee in ditches and low muddy places and on Pelee island, Kelley, and North Bass islands, and Ohio shore.

Eleocharis palustris (L.) R. and S. (CREEPING SPIKE RUSH.)
Common at Point Pelee and on Pelee island.
Probably throughout Essex county. Ohio shore.

Eleocharis palustris (L.) R. and S., var. **glaucescens** (Willd.) Gray. (SLENDER CREEPING SPIKE RUSH.)
On Pelee island, Put-in-Bay island, and Ohio shore.

Eleocharis palustris (L.) R. and S., var. **vigens** Bailey. (LARGE CREEPING SPIKE RUSH.)
In water along Ohio shore.

Eleocharis acicularis (L.) R. and S. (NEEDLE SPIKE RUSH.)
Common at Point Pelee about the ponds and small lakes of the big marsh and on Pelee island, Bass island, and Ohio shore.

Eleocharis tenuis (Willd.) Schultes. (SLENDER SPIKE RUSH.)
Occasional at Point Pelee in damp grassy ground, and on Pelee island.

Eleocharis acuminata (Muhl.) Nees. (FLAT-STEMMED SPIKE RUSH.)
Common on Ohio shore.

Eleocharis intermedia (Muhl.) Schultes. (MATTED SPIKE RUSH.)
Bass island and Ohio shore.

Stenophyllus capillaris (L.) Britton. (HAIR-LIKE STENOPHYLLUS.)
In wet sandy fields near Sandwich. (Macoun.)

Fimbristylis autumnalis (L.) R. and S. (SLENDER FIM-
BRISTYLIS.)

Wet sandy fields near Sandwich. (Macoun.) Ohio shore.

Scirpus debilis Pursh. (WEAK-STALKED CLUB RUSH.)

Along Ohio shore.

Scirpus americanus Pers. (THREE-SQUARE.)

Very common at Point Pelee on borders of big marsh,
and along shore of Lake St. Clair. Islands and Ohio
shore.

Scirpus Torreyi Olney. (TORREY'S BULRUSH.)

Along Ohio shore.

Scirpus occidentalis (Wats.) Chase. (WESTERN BULRUSH.)

Common at Point Pelee about and in the big marsh, and
along Lake St. Clair shore. Islands and Ohio shore.

Scirpus fluviatilis (Torr.) Gray. (RIVER BULRUSH.)

Frequent at Point Pelee in the very wet places of the big
marsh. Put-in-Bay island and Ohio shore.

Scirpus sylvaticus L. (WOOD BULRUSH.)

Put-in-Bay island.

Scirpus atrovirens Muhl. (DARK GREEN BULRUSH.)

Common at Point Pelee in damp open ground and
throughout Essex county. Islands and Ohio shore.

Scirpus polyphyllus Vahl. (LEAFY BULRUSH.)

Middle Bass island.

Scirpus lineatus Michx. (REDDISH BULRUSH.)

In damp open meadow-like ground on the east side of
Point Pelee. Apparently infrequent. Kelley and
North Bass islands and Ohio shore.

Scirpus cyperinus (L.) Kunth., var. **pelius** Fernald. (WOOD
GRASS.)

Occasional in marshy places near Lake St. Clair. Along
Ohio shore.

Hemicarpha micrantha (Vahl.) Britton. (HEMICARPHA.)

In damp sandy ground near Amherstburg. (Macoun.)

Cladium mariscoides (Muhl). Torr. (TWIG RUSH.)

Frequent and in spots abundant at Point Pelee in wet
marshy places and about the ponds and small lakes
of the big marsh.

Scleria triglomerata Michx. (TALL NUT RUSH.)

In damp open ground near Sandwich. (Macoun.)

Carex muskingumensis Schwein. (MUSKINGUM SEDGE.)

Along Ohio shore,

Carex scoparia Schkuhr. (POINTED BROOM SEDGE.)

Frequent at Point Pelee in wet open places, and on Pelee island.

Carex tribuloides Wahlenb. (BLUNT BROOM SEDGE.)

Occasional at Point Pelee in damp open places. North Bass island and Ohio shore.

Carex tribuloides Wahlenb., var. **turbata** Bailey. (BLUNT BROOM SEDGE.)

North Bass island.

Carex cristata Schwein. (CRESTED SEDGE.)

In wet open ground and damp open woods at Point Pelee. North Bass island and Ohio shore.

Carex festucacea Schkuhr., var. **brevior** (Dewey) Fernald. (FESCUE SEDGE.)

Johnson, Kelley, and Green islands and Ohio shore.

Carex Bebbii Olney. (BEBB'S SEDGE.)

In low damp open ground at Point Pelee. Apparently infrequent.

Carex sterilis Willd. (LITTLE PRICKLY SEDGE.)

Ohio shore.

Carex rosea Schkuhr. (STELLATE SEDGE.)

Frequent at Point Pelee in dry open woods. Middle Bass island and Ohio shore.

Carex rosea Schkuhr., var **radiata** Dewey. (STELLATE SEDGE.)

In woods near Amherstburg. (Macoun.)

Carex Muhlenbergii Schkuhr. (MUHLENBERG'S SEDGE.)

Occasional at Point Pelee on open sandy ground. Ohio shore.

Carex Muhlenbergii Schkuhr., var. **enervis** Boott. (MUHLENBERG'S SEDGE.)

Ohio shore.

Carex cephalophora Muhl. (OVAL-HEADED SEDGE.)

Occasional at Point Pelee in damp shaded ground. Bass island and Ohio shore.

Carex cephaloidea Dewey. (THIN-LEAVED SEDGE.)
Ohio shore.

Carex sparganioides Muhl. (BUR REED SEDGE.)
In shaded ground on Pelee island. Kelley and Rattlesnake islands and Ohio shore.

Carex vul· noidea Michx. (FOX SEDGE.)
Common at Point Pelee and throughout Essex county
in damp meadow-like ground Islands and Ohio
shore.

Carex stipata Muhl. (AWL-FRUITED SEDGE.)
In very wet open places at Point Pelee, especially on
borders of big marsh. Islands and Ohio shore.

Carex Sartwellii Dewey. (SARTWELL'S SEDGE.)
Wet open places in the big marsh at Point Pelee.

Carex aquatilis Wahlenb. (WATER SEDGE.)
Margins of ponds and small lakes in the big marsh at
Point Pelee. Put-in-Bay island and Ohio shore.

Carex torta Boott. (TWISTED SEDGE.)
On Cedar point, Ohio shore.

Carex stricta Lam. (TUSSOCK SEDGE.)
About the big marsh at Point Pelee, especially near the
ponds and small lakes.

Carex leptalea Wahlenb. (BRISTLE-STALKED SEDGE.)
Common at Point Pelee in very wet open places. Ohio
shore.

Carex virescens Muhl., var. **Swanii** Fernald. (SWAN'S SEDGE.)
Near Essex Centre and Amherstburg. (Macoun.)
Near Leamington. (Burgess.)

Carex Davisii Schwein. (DAVIS' SEDGE.)
Kelley island and Ohio shore.

Carex albicans Willd. (WHITISH SEDGE.)
Put-in-Bay island.

Carex varia Muhl. (EMMONS' SEDGE.)
Occasional at Point Pelee in dry sandy ground among
red cedars and pines. Ohio shore.

Carex pennsylvanica Lam. (PENNSYLVANIA SEDGE.)
Frequent at Point Pelee on dry open ground. Put-in-
Bay island and Ohio shore.

Carex eburnea Boott. (BRISTLE-LEAVED SEDGE.)

On dry sandy ground at Point Pelee among red cedars and pines.

Carex laxiflora Lam. (LOOSE-FLOWERED SEDGE.)

Near Amherstburg. (Macoun.) Kelley island.

Carex laxiflora Lam., var. **varians** Bailey. (LOOSE-FLOWERED SEDGE.)

Occasional at Point Pelee in shaded ground. Hen and Kelley islands, and Ohio shore.

Carex laxiflora Lam., var. **blanda** (Dewey) Boott. (LOOSE-FLOWERED SEDGE.)

Frequent at Point Pelee in rich woods. Kelley island and Ohio shore.

Carex laxiflora Lam., var. **latifolia** Boott. (WHITE BEAR SEDGE.)

Kelley island.

Carex granularis Muhl. (MEADOW SEDGE.)

Kelley island and Ohio shore.

Carex Crawei Dewey. (CRAWE'S SEDGE.)

Ohio shore.

Carex flava L. (YELLOW SEDGE.)

Occasional at Point Pelee in wet open ground.

Carex Oederi Retz., var. **pumila** (Cosson and Germain) Fernald. (GREEN SEDGE.)

In damp sand about the ponds and small lakes of the big marsh at Point Pelee, and along the north shore of Lake Erie.

Carex filiformis L. (SLENDER SEDGE.)

Common at Point Pelee in and about the big marsh, and on Pelee island.

Carex lanuginosa Michx. (WOOLLY SEDGE.)

In damp meadow-like ground at Point Pelee. Put-in-Bay island.

Carex riparia W. Curtis. (RIVER-BANK SEDGE.)

On marshes at Point Pelee. Ohio shore.

Carex squarrosa L. (SQUARROSE SEDGE.)

Ohio shore.

Carex Frankii Kunth. (FRANK'S SEDGE.)
Middle Bass island and Ohio shore.
Carex comosa Boott. (BRISTLY SEDGE.)
Common at Point Pelee in wet spots on the big marsh.
Islands.
Carex hystericina Muhl. (PORCUPINE SEDGE.)
Common at Point Pelee in very wet places, especially
on the big marsh. Put-in-Bay and Middle Bass
islands and Ohio shore.
Carex lurida Wahlenb. (SALLOW SEDGE.)
Ohio shore
Carex lupulina Muhl. (HOP SEDGE.)
Common at Point Pelee in wet marshy spots. Islands
and Ohio shore.
Carex intumescens Rudge. (BLADDER SEDGE.)
Near Amherstburg. (Macoun.)

ARACEAE (ARUM FAMILY.)

Arisaema triphyllum (L.) Schott. (JACK-IN-THE-PULPIT.)
Common at Point Pelee in rich woods. Islands and Ohio
shore.
Arisaema Dracontium (L.) Schott. (GREEN DRAGON.)
Frequent on Grosse Isle, Wayne county, Mich., and
should be found in Essex county. Ohio shore.
Symplocarpus foetidus (L.) Nutt. (SKUNK CABBAGE.)
Common about Windsor.
Acorus Calamus L. (SWEET FLAG.)
Common at Point Pelee on border of the big marsh,
especially along "the narrows," and west of Colchester.
Put-in-Bay and Ohio shore.

LEMNACEAE (DUCKWEED FAMILY.)

Spirodela polyrhiza (L.) Schleid. (GREATER DUCKWEED.)
Common at Point Pelee on water of ditches and pools
in the big marsh. Ohio shore.

Lemna trisulca L. (Ivy-leaved Duckweed.)

    On and in water about Put-in-Bay island, and Ohio shore.

Lemna minor L. (Lesser Duckweed.)

    Often covering stagnant water in the big marsh at Point Pelee, also about the islands and Ohio shore.

Wolffia columbiana Karst. (Columbia Wolffia.)

    About the Ohio shore.

Wolffia punctata Griseb. (Brazil Wolffia.)

    Ohio shore.

COMMELINACEAE (Spiderwort Family.)

Tradescantia reflexa Raf. (Reflexed Spiderwort.)

    Ohio shore.

Tradescantia virginiana L. (Spiderwort.)

    Frequent on Ohio shore.

PONTEDERIACEAE (Pickerel Weed Family.)

Pontederia cordata L. (Pickerel Weed.)

    Frequent at Point Pelee in shallow water of the ponds and small lakes of the big marsh. Islands and Ohio shore.

Heteranthera dubia (Jacq.) MacM. (Water Star Grass.)

    In still water about the islands and Ohio shore.

JUNCACEAE (Rush Family.)

Juncus bufonius L. (Toad Rush.)

    Common at Point Pelee in damp open ground and on Point Pelee island.

Juncus tenuis Willd. (Slender Rush.)

    Common at Point Pelee about and near dwellings, along roads and paths. Islands and Ohio shore.

Juncus balticus Willd., var. **littoralis** Engelm. (BALTIC RUSH.)

>   Occasional at Point Pelee in open damp or dry ground. Not noticed here on the beach as a sand binder. Ohio shore.

Juncus effusus L. (COMMON RUSH.)

>   Occasional at Point Pelee on open damp ground. North Bass island and Ohio shore.

Juncus brachycephalus (Engelm.) Buchenau. (SMALL-HEADED RUSH.)

>   Ohio shore.

Juncus canadensis J. Gay. (CANADA RUSH.)

>   Occasional at Point Pelee in marshy ground. Ohio shore.

Juncus nodosus L. (KNOTTED RUSH.)

>   Frequent at Point Pelee on borders of big marsh and in damp sand about the ponds and small lakes. In ditches near Leamington. (Burgess.) Ohio shore.

Juncus acuminatus Michx. (SHARP-FRUITED RUSH.)

>   Near Essex Centre and at Point Pelee. (Macoun.)

Juncus Torreyi Coville. (TORREY'S RUSH.)

>   Frequent at Point Pelee in low damp ground. Ohio shore.

Juncus alpinus Vill., var. **insignis** Fries. (RICHARDSON'S RUSH.)

>   Occasional at Point Pelee on the sandy beach. Kelley island and Ohio shore.

Luzula saltuensis Fernald. (HAIRY WOOD RUSH.)

>   Frequent at Point Pelee in open woods and thickets.

.zula campestris (L.) DC., var. **multiflora** (Ehrh.) Celak. (COMMON WOOD RUSH.)

>   Occasional at Point Pelee in open dry ground.

LILIACEAE (LILY FAMILY.)

Zygadenus chloranthus Richards. (GLAUCOUS ZYGADENUS.)

>   Ohio shore.

Uvularia grandiflora Sm. (LARGE-FLOWERED BELLWORT.)
>   Frequent at Point Pelee in damp rich woods. Islands
>   and Ohio shore.

Oakesia sessilifolia (L.) Wats. (SESSILE-LEAVED BELLWORT.)
>   Common about Windsor.

Allium tricoccum Ait. (WILD LEEK.)
>   Point Pelee island. Occasional in other parts of Essex
>   county, and formerly abundant.

Allium cernuum Roth. (WILD ONION.)
>   Pelee island, Newton Tripp, 1913. Islands and Ohio
>   shore.

Allium canadense L. (WILD GARLIC.)
>   Abundant at Point Pelee in spots. Near Colchester.
>   (Macoun.) Kelley island and Ohio shore.

Hemerocallis fulva L. (COMMON DAY LILY.)
>   A common escape from cultivation and well established
>   near dwellings at Point Pelee. North Bass island
>   and Ohio shore.

Lilium philadelphicum L. (WILD ORANGE-RED LILY.)
>   Ohio shore.

Lilium philadelphicum L., var. **andinum** (Nutt.) Ker.
(WESTERN RED LILY.)
>   In open sandy places along shore of Lake St. Clair.

Lilium canadense L. (WILD YELLOW LILY.)
>   Frequent at Point Pelee and on Pelee island. Kelley
>   island.

Erythronium americanum Ker. (YELLOW ADDER'S TONGUE.)
>   Common in woods about Windsor. Islands and Ohio.
>   shore.

Erythronium albidum Nutt. (WHITE ADDER'S TONGUE.)
>   In rich woods about Windsor. Johnson, Kelley, and
>   Rattlesnake islands and Ohio shore.

Camassia esculenta (Ker) Robinson. (WILD HYACINTH.)
>   White island in Detroit river. (Macoun.) On eight
>   islands, and Ohio shore.

Ornithogalum umbellatum L. (STAR OF BETHLEHEM.)
>   Put-in-Bay island.

Asparagus officinalis L. (GARDEN ASPARAGUS.)

Well established at Point Pelee in open light sandy ground. Islands and Ohio shore.

Smilacina racemosa (L.) Desf. (FALSE SPIKENARD.)

Common at Point Pelee in rich open woods or in open or slightly shaded ground. Common on the islands and Ohio shore.

Smilacina stellata (L.) Desf. (STAR-FLOWERED SOLOMON'S SEAL.)

Common at Point Pelee in rich open woods or in sandy open ground on the upper beach. Common on the islands and Ohio shore.

Smilacina trifolia (L.) Desf. (THREE-LEAVED SOLOMON'S SEAL.)

Frequent near Windsor in very wet places (F. P. Cravin.)

Maianthemum canadense Desf. (FALSE LILY-OF-THE-VALLEY.)

Common at Point Pelee in dry woods. Ohio shore.

Streptopus roseus Michx. (SESSILE-LEAVED TWISTED STALK.)

In thick damp woods about Windsor. (F. P. Cravin).

Polygonatum biflorum (Walt.) Ell. (SMALL SOLOMON'S SEAL.)

Common at Point Pelee in rich open woods and thickets. Ohio shore.

Polygonatum commutatum (R. and S.) Dietr. (GREAT SOLOMON'S Seal.)

Common on north shore of Lake Erie and along Detroit river. Common on the islands and Ohio shore.

Trillium erectum L. (ILL-SCENTED WAKE ROBIN.)

Frequent about Windsor. (F. P. Cravin.) Common on the islands and Ohio shore.

Trillium grandiflorum (Michx.) Salisb. (LARGE-FLOWERED WAKE ROBIN.)

Frequent at Point Pelee in rich open woods and thickets. Common on the islands and Ohio shore.

Aletris farinosa L. (COLIC-ROOT.)

In sandy thickets near Leamington. (Burgess.) Common near Sandwich. (Macoun.)

Smilax herbacea L. (CARRION-FLOWER.)
> Frequent at Point Pelee in open woods and on Pelee island and throughout Essex county. Ohio shore.

Smilax ecirrhata (Engelm.) Wats. (UPRIGHT SMILAX.)
> Frequent at Point Pelee in rich woods and thickets. Kelley island and Ohio shore.

Smilax rotundifolia L. (COMMON GREEN BRIER.)
> Put-in-Bay island and Ohio shore.

Smilax rotundifolia L., var. **quadrangularis** (Muhl.) Wood. (SQUARE-STEMMED GREEN BRIER.)
> In damp woods at Point Pelee. (Macoun.) Low woods near Leamington. (Burgess.)

Smilax hispida Muhl. (HISPID GREEN BRIER.)
> Common at Point Pelee in moist thickets. Islands and Ohio shore.

DIOSCOREACEAE (YAM FAMILY.)

Dioscorea villosa L. (WILD YAM-ROOT.)
> Plentiful at Point Pelee in woods and thickets and on Pelee island. Ohio shore. (*D. paniculata* Michx. according to Bulletin 89, United States Department of Agriculture, Bureau of Plant Industry.)

AMARYLLIDACEAE (AMARYLLIS FAMILY.)

Hypoxis hirsuta (L.) Coville.
> Common in meadow-like ground about Windsor. (F. P. Cravin.) Ohio shore.

IRIDACEAE (IRIS FAMILY.)

Iris versicolor L. (LARGE BLUE FLAG.)
> Borders of big marsh at Point Pelee and other damp open places. Islands and Ohio shore.

Sisyrinchium angustifolium Mill. (NORTHERN BLUE-EYED GRASS.)
> Frequent about Windsor. Ohio shore.

Sisyrinchium gramineum Curtis. (COMMON BLUE-EYED GRASS.)

> Occasional about Windsor. Ohio shore.

ORCHIDACEAE (ORCHIS FAMILY.)

Cypripedium parviflorum Salisb., var. **pubescens** (Willd.) Knight. (LARGER YELLOW LADY'S SLIPPER.)

> Ohio shore.

Habenaria flava (L.) Gray. (TUBERCLED ORCHIS.)

> On Cedar point, Ohio shore.

Habenaria ciliaris (L.) R. Br. (YELLOW FRINGED ORCHIS.)

> In sandy ground near Leamington. (Burgess.) Near Windsor. (F. P. Cravin.)

Habenaria psycodes (L.) Sw. (SMALLER PURPLE-FRINGED ORCHIS.)

> Cedar point, Ohio shore.

Pogonia ophioglossoides (L.) Ker. (ROSE POGONIA.)

> Frequent in damp meadow-like ground about Windsor. (F. P. Cravin.)

Spiranthes cernua (L.) Richard. (NODDING LADIES' TRESSES.)

> Common along north shore of Lake Erie.

Epipactis pubescens (Willd.) A. A. Eaton. (DOWNY RATTLE-SNAKE PLANTAIN.)

> In rich woods near Windsor. (F. P. Cravin.)

Corallorrhiza maculata Raf. (LARGE CORAL ROOT.)

> Ohio shore.

Liparis Loeselii (L.) Richard. (FEN ORCHIS.)

> Cedar point, Ohio shore.

Aplectrum hyemale (Muhl.) Torr. (PUTTY-ROOT.)

> Ohio shore.

PIPERACEAE (PEPPER FAMILY.)

Saururus cernuus L. (LIZARD'S TAIL.)

> In wet places near Detroit river below Amherstburg. (Maclagan.)

SALICACEAE (WILLOW FAMILY.)

Salix nigra Marsh. (BLACK WILLOW.)
>Frequent at Point Pelee in damp places on borders of woods and occasional on the beach. Usually mere shrubs. Islands and Ohio shore

Salix amygdaloides Anders. (PEACH-LEAVED WILLOW.)
>Frequent at Point Pelee in damp open ground, and occasional along "the narrows". Abundant on Pelee island.

Salix pentandra L. (BAY-LEAVED WILLOW.)
>Occasionally planted, but not noticed as escaping.

Salix lucida Muhl. (SHINING WILLOW.)
>Occasional at north end of Point Pelee. Hen and Put-in-Bay islands and Ohio shore.

Salix alba L. (WHITE WILLOW.)
>Often planted and apparently spreading at Point Pelee and on Pelee island.

Salix alba L., var. **vitellina** (L.) Koch. (WHITE WILLOW.)
>Islands and Ohio shore.

Salix babylonica L. (WEEPING WILLOW.)
>Planted in abundance and thriving along roads near Lake St. Clair. Not noticed as spreading.

Salix longifolia Muhl. (SAND BAR WILLOW.)
>Occasional at Point Pelee in damp open ground near the big marsh and in damp sand on and near the beach. Common on the islands and Ohio shore.

Salix cordata Muhl. (HEART-LEAVED WILLOW.)
>Frequent on Pelee island and the other islands and Ohio shore.

Salix glaucophylla Bebb. (BROAD-LEAVED WILLOW.)
>On Cedar point, Ohio shore.

Salix balsamifera Barratt. (BALSAM WILLOW.)
>Reported from Kent county, and should be found in Essex county.

Salix pedicellaris Pursh. (BAG WILLOW.)
>In very wet places about Windsor.

Salix discolor Muhl. (GLAUCOUS WILLOW.)
> Occasional on Pelee island. Hen island and Ohio shore.

Salix discolor Muhl., var. **eriocephala** (Michx.) Anders.
(GLAUCOUS WILLOW.)
> Ohio shore.

Salix petiolaris Sm.
> Damp ground about Windsor.

Salix humilis Marsh. (PRAIRIE WILLOW.)
> In dry open ground about Windsor.

Salix sericea Marsh. (SILKY WILLOW.)
> Swampy places near Windsor.

Salix rostrata Richards. (BEBB'S WILLOW.)
> Frequent on Pelee island. Other islands and Ohio shore.

Salix candida Flugge. (SAGE WILLOW.)
> Open marshy ground near Windsor.

Salix purpurea L. (PURPLE WILLOW.)
> In damp open ground at Point Pelee near Grubb's fishery
> buildings. Kelley and Put-in-Bay islands and Ohio
> shore.

Populus alba L. (WHITE POPLAR.)
> Occasional at Point Pelee as a cultivated tree and spread-
> ing by root. Kelley and Put-in-Bay islands.

Populus tremuloides Michx. (AMERICAN ASPEN.)
> Frequent at Point Pelee but nowhere abundant. Islands.

Populus grandidentata Michx. (LARGE-TOOTHED ASPEN.)
> Occasional at Point Pelee and on Pelee island. Put-in-
> Bay island and Ohio shore.

Populus balsamifera L. (BALSAM POPLAR.)
> Small trees noticed at Point Pelee in dry open ground
> and along the west upper beach. Ohio shore.

Populus deltoides Marsh. (COTTON-WOOD.)
> Many large trees along "the narrows" and in rich ground
> with other trees at Point Pelee. Common on the
> islands and Ohio shore.

Populus nigra L., var. **italica** Du Roi. (LOMBARDY POPLAR.)
> Occasionally planted at Point Pelee and spreading by
> root.

JUGLANDACEAE (Walnut Family.)

Juglans cinera L. (Butternut.)

Occasional at Point Pelee in dry ground with other trees. Frequent along the north shore of Lake Erie. Ohio shore.

Juglans nigra L. (Black Walnur.)

Abundant at Point Pelee growing in pure sand, especially along "the narrows". Very common along the north shore of Lake Erie and on Pelee island. Formerly on Kelley and Middle Bass islands. Ohio shore.

Carya ovata (Mill.) K. Koch. (*C. Alba* Nutt.) (Shag-bark Hickory.)

Occasional at Point Pelee and near Lake St. Clair. Abundant on the islands and Ohio shore.

Carya laciniosa (Michx. f.) Loud. (*C. sulcata* Nutt.) (Big Shell-bark.)

Near Colchester and probably throughout Essex county. Ohio shore.

Carya alba (L.) K. Koch. (*C. tomentosa* Nutt.) (Mocker Nut.)

Very common at Point Pelee and on Pelee island. Put-in-Bay island.

Carya microcarpa Nutt. (Small-fruited Hickory.)

Plentiful along north shore of Lake Erie, and probably throughout Essex county. Pelee island.

Carya glabra (Mill.) Spach. (Pignut Hickory.)

Near Amherstburg. Islands and Ohio shore.

Carya cordifromis (Wang.) K. Koch. (*C. amara* Nutt.) (Bitternut.)

Frequent at Point Pelee and on Pelee island. Ohio shore.

BETULACEAE (Birch Family.)

Corylus americana Walt. (Common Hazelnut.)

Occasional at Point Pelee on borders of woods and on Pelee island. Not noticed on the other islands. Ohio shore.

Ostrya virginiana (Mill.) K. Koch. (IRONWOOD.)

  Common at Point Pelee in rich ground with other trees. Islands and Ohio shore.

Carpinus caroliniana Walt. (BLUE BEECH.)

  Infrequent at Point Pelee, but common along north shore of Lake Erie. Probably throughout Essex county. Along Lake St. Clair. Formerly on Kelley island.

Betula lutea Michx. f. (YELLOW BIRCH.)

  Near Kingsville and farther west along north shore of Lake Erie.

Alnus incana (L.) Moench. (SPECKLED ALDER.)

  In damp places near Lake St. Clair. Apparently infrequent.

FAGACEAE (BEECH FAMILY.)

Fagus grandifolia Ehrh. (BEECH.)

  Not noticed at Point Pelee, but common along the north shore of Lake Erie to Detroit river. Formerly on Put-in-Bay and Middle Bass islands.

Castanea dentata (Marsh.) Borkh. (CHESTNUT.)

  Frequent along the north shore of Lake Erie to Detroit river. Fine specimens still existing in woods about Windsor.

Quercus alba L. (WHITE OAK.)

  Not noticed at Point Pelee, but common on Pelee island. No doubt throughout Essex county. On the other islands and Ohio shore.

Quercus macrocarpa Michx. (BUR OAK.)

  Frequent at Point Pelee in rich ground with other trees. Islands and Ohio shore.

Quercus bicolor Willd. (SWAMP WHITE OAK.)

  Occasional at Point Pelee in rich ground with other trees and on Pelee island, Lake Erie shore, Kelley island, and Ohio shore.

Quercus Muhlenbergii Engelm. (CHESTNUT OAK.)

  Frequent at Point Pelee with other trees, and along Lake Erie shore. Islands and Ohio shore.

Quercus prinoides Willd. (SCRUB CHESTNUT OAK.)

Noticed by Thomas Burgess at Point Pelee.

Quercus rubra L. (RED OAK.)

Very common at Point Pelee with other trees in both dry and damp rich ground. Common on islands and Ohio shore.

Quercus palustris Muench. (PIN OAK.)

Near Leamington (Burgess.) Ohio shore. Fine trees about Windsor.

Quercus coccinea Muench. (SCARLET OAK.)

Frequent in dry ground about Windsor. (F. P. Cravin). Ohio shore.

Quercus velutina Lam. (YELLOW OAK.)

Frequent at Point Pelee in dry ground with other trees. Kelley and Put-in-Bay islands and Ohio shore.

Quercus imbricaria Michx. (LAUREL OAK.)

Abundant on Cedar point, Ohio shore.

URTICACEAE (NETTLE FAMILY.)

Ulmus fulva Michx. (SLIPPERY ELM.)

Occasional at Point Pelee in rich ground with other trees. Very abundant on Pelee island. The other islands and Ohio shore.

Ulmus americana L. (AMERICAN ELM.)

Abundant at Point Pelee in rich ground with other trees. Common on the islands and Ohio shore.

Ulmus racemosa Thomas. (ROCK ELM.)

Frequent about Windsor. (F. P. Cravin.)

Celtis occidentalis L. (SUGARBERRY.)

Abundant in sandy ground at Point Pelee, especially along "the narrows." Lake Erie shore and along Detroit river. Common on the islands and Ohio shore.

Cannabis sativa L. (HEMP.)

Very abundant at Point Pelee in spots in dry shaded ground growing like a native plant. Point Pelee island.

Humulus Lupulus L. (COMMON HOP.)

Noticed as an escape at Point Pelee and on Point Pelee island.

Maclura pomifera (Rat.) Schneider. (OSAGE ORANGE.)

Often planted for hedges in Essex county, but perhaps not spreading as an escape.

Morus rubra L. (RED MULBERRY.)

Said to be plentiful formerly at Point Pelee, but a few small trees only were noticed in 1911. Abundant, and trees often large on Pelee island. Frequent along the north shore of Lake Erie. On the other islands and Ohio shore.

Urtica gracilis Ait. (SLENDER NETTLE.)

Common at Point Pelee on borders of the big marsh along "the narrows" and other damp open or shaded places. Common on the islands and Ohio shore.

Laportea canadensis (L). Gaud. (WOOD NETTLE.)

Common at Point Pelee in rich woods. Also common on the islands and Ohio shore.

Pilea pumila (L.) Gray. (RICHWEED.)

Abundant in spots at Point Pelee in damp shaded ground. Kelley island and Ohio shore.

Boehmeria cylindrica (L.) Sw. (WILD FALSE NETTLE.)

Frequent at Point Pelee in damp open or slightly shaded places. Common on the islands and Ohio shore.

Parietaria pennsylvanica Muhl. (PENNSYLVANIA PELLITORY.)

Abundant at Point Pelee under red cedars and other trees. Abundant on the islands and Ohio shore.

SANTALACEAE (SANDALWOOD FAMILY.)

Comandra umbellata (L.) Nutt. (BASTARD TOAD-FLAX.)

Frequent at Point Pelee in open or shaded sandy ground and on Pelee island. Ohio shore.

ARISTOLOCHIACEAE (Birthwort Family.)

Asarum canadense L. (Wild Ginger.)
In rich shaded ground along Detroit river. (Maclagan.)

POLYGONACEAE (Buckwheat Family.)

Rumex Britannica L. (Great Water Dock.)
Frequent at Point Pelee in and about the big marsh, and on Pelee island. Cedar point, Ohio shore.

Rumex crispus L. (Yellow Dock.)
Occasional at Point Pelee about dwellings and in old fields. Abundant on the islands and Ohio shore.

Rumex altissimus Wood. (Pale Dock.)
Put-in-Bay island.

Rumex verticillatus L. (Swamp Dock.)
Occasional at Point Pelee in very wet open places. Ohio shore.

Rumex obtusifolius L. (Bitter Dock.)
About dwellings and in old cultivated fields at Point Pelee. Common on the islands and Ohio shore.

Rumex Acetosella L. (Field Sorrel.)
Occasional at Point Pelee near dwellings and in old fields. Put-in-Bay and Kelley islands and Ohio shore.

Polygonum aviculare L. (Knot Grass.)
Common at Point Pelee about dwellings and in waste places. Abundant on the islands and Ohio shore.

Polygonum aviculare L., var. **littorale** (Link) Koch. (Shore Knotweed.)
Kelley island.

Polygonum erectum L. (Erect Knotweed.)
Occasional about dwellings at Point Pelee. Common on the islands and Ohio shore.

Polygonum tenue Michx. (Slender Knotweed.)
Ohio shore.

Polygonum lapathifolium L. (DOCK-LEAVED PERSICARIA.)

Common at Point Pelee on the banks of big ditches and in other wet places of the big marsh, and on Pelee island. Ohio shore.

Polygonum tomentosum Schrank. (SLENDER PINK PERSICARIA.)

Frequent at Point Pelee along borders of the big marsh, and occasional as a weed near dwellings. Ohio shore.

Polygonum amphibium L. (WATER PERSICARIA.)

Frequent at Point Pelee along ditches and in wet places on and about the big marsh, and on Pelee island. Ohio shore.

Polygonum Muhlenbergii (Meisn.) Wats. (SWAMP PERSICARIA.)

Common at Point Pelee in very wet places on the big marsh. Islands and Ohio shore.

Polygonum pennsylvanicum L. (PENNSYLVANIA PERSICARIA.)

Frequent at Point Pelee in damp meadow-like ground. Along Detroit river. (Maclagan.) Kelley and Middle Bass islands and Ohio shore.

Polygonum Hydropiper L. (COMMON SMARTWEED.)

Frequent at Point Pelee in damp open or slightly shaded ground. Common on the islands and Ohio shore.

Polygonun acre HBK. (WATER SMARTWEED.)

Abundant at Point Pelee bordering the big marsh along "the narrows" and in other damp places. Islands and Ohio shore.

Polygonum orientale L. (PRINCE'S FEATHER.)

Noticed at Point Pelee as an escape near dwellings, and apparently persisting.

Polygonum Persicaria L. (LADY'S THUMB.)

Occasional at Point Pelee about dwellings and often abundant on banks of ditches. Abundant on the islands and Ohio shore.

Polygonum hydropiperoides Michx. (WILD WATER PEPPER.)

Occasional at Point Pelee in wet shaded places. Kelley island.

Polygonum virginianum L. (Virginia Knotweed.)
> In rich woods at Point Pelee. Ohio shore.

Polygonum sagittatum L. (Arrow-leaved Tear-thumb.)
> Occasional at Point Pelee in low open or shaded ground and in damp places along north shore of Lake Erie. Ohio shore.

Polygonum Convolvulus L. (Black Binweed.)
> Frequent at Point Pelee as a weed in gardens and cultivated fields. Common on the islands and Ohio shore.

Polygonum scandens L. (Climbing False Buckwheat.)
> Frequent at Point Pelee in rich shaded ground and on Pelee island. The other islands and Ohio shore.

Polygonum dumetorum L. (Copse Buckwheat.)
> Ohio shore.

Fagopyrum esculentum Moench. (Buckwheat.)
> Occasional at Point Pelee as an escape. Ohio shore.

CHENOPODIACEAE (Goosefoot Family.)

Cycloloma atriplicifolium (Spreng.) Coult. (Winged Pigweed.)
> In sandy ground near Windsor. Cedar point, Ohio shore.

Chenopodium ambrosioides L. (Mexican Tea.)
> In and about cultivated grounds along Detroit river.

Chenopodium Botrys L. (Jerusalem Oak.)
> Occasional at Point Pelee in dry sandy ground. Kelley island and Ohio shore.

Chenopodium capitatum (L.) Asch. (Strawberry Blite.)
> Middle Bass and Green islands.

Chenopodium hybridum L. (Maple-leaved Goosefoot.)
> Noticed at Point Pelee as a weed in gardens and cultivated fields. Islands and Ohio shore.

Chenopodium album L. (Lamb's Quarters.)
> Common at Point Pelee as a weed in gardens, cultivated fields, and waste places. Common on the islands and Ohio shore.

Chenopodium album L., var. **viride** (L.) Moq. (Pigweed.)
> Common on the islands and Ohio shore.

Chenopodium urbicum L. (Upright Goosefoot.)
> Kelley island.

Chenopodium Boscianum Moq. (Basc's Goosefoot.)
> In sandy thickets near Point Pelee. (Macoun.) Kelley island.

Chenopodium leptophyllum Nutt. (Narrow-leaved Goosefoot.)
> Sandy woodlands at Point Pelee. (Burgess.)

Atriplex patula L., var. **hastata** (L'Gray. (Halberd-leaved Orache.)
> Occasional at Point Pelee as a weed about dwellings and on Pelee island. Ohio shore.

Salsola Kali L., var. **tenuifolia** G.F.W. Mey. (Russian Thistle.)
> Occasional at Point Pelee as a weed about dwellings.

AMARANTHACEAE (Amaranth Family.)

Amaranthus retroflexus L. (Amaranth Pigweed.)
> A common weed at Point Pelee in gardens and cultivated fields. Common on the islands and Ohio shore.

Amaranthus hydridus L. (Green Amaranth.)
> Noticed near Leamington by Dearness Common on the islands.

Amaranthus paniculatus L. (Purple Amaranth.)
> Islands.

Amaranthus graecizens L. (Tumble Weed.)
> Occasional at Point Pelee as a weed in sandy cultivated fields. Islands and Ohio shore.

Amaranthus blitoides Wats. (Prostrate Amaranth.)
> Occasional at Point Pelee as a weed about dwellings and in gardens. Islands and Ohio shore.

Acnida tuberculata Moq. (Western Water Hemp.)
> On Kelley and Middle Bass islands and Ohio shore.

PHYTOLACCACEAE (POKEWEED FAMILY.)

Phytolacca decandra L. (COMMON POKE.)
> Common at Point Pelee in shaded sandy ground. Along Detroit river. (Maclagan.) Islands and Ohio shore.

NYCTAGINACEAE (FOUR-O'CLOCK FAMILY.)

Oxybaphus nyctagineus (Michx.) Sweet. (HEART-LEAVED UMBRELLA-WORT.)
> Along railways near Windsor. Cedar point, Ohio shore.

ILLECEBRACEAE (KNOTWORT FAMILY.)

Anychia polygonoides Raf. (SLENDER FORKED CHICKWEED.)
> Put-in-Bay island.

Anychia canadensis (L.) BSP. (FORKED CHICKWEED.)
> Near Amherstburg. (Macoun.) Ohio shore.

AIZOACEAE (CARPET WEED FAMILY.)

Mollugo verticillata L. (CARPET WEED.)
> Occasional at Point Pelee in open sandy ground and on Pelee island.

CARYOPHYLLACEAE (PINK FAMILY.)

Arenaria lateriflora L. (BLUNT-LEAVED SANDWORT.)
> Ohio shore.

Arenaria serpyllifolia L. (THYME-LEAVED SANDWORT.)
> Frequent at Point Pelee in dry open ground near dwellings. Islands and Ohio shore.

Arenaria stricta Michx. (ROCK SANDWORT.)
> Occasional at Point Pelee in open sandy ground, and often on the upper beach. Islands and Ohio shore.

Stellaria longifolia Muhl. (LONG-LEAVED STITCHWORT.)
>Frequent at Point Pelee in open damp grassy places and open damp woods, and abundant in spots on the big marsh. Pelee island. Ohio shore.

Stellaria media (L.) Cyrill. (COMMON CHICKWEED.)
>Common at Point Pelee as a garden and field weed. Abundant on the islands and Ohio shore.

Cerastium arvense L., var. **oblongifolium** (Torr.) Hollick and Britton. (FIELD MOUSE-EAR CHICKWEED.)
>In sandy ground at Point Pelee and on Pelee island. Near Amherstburg. The other islands and Ohio shore.

Cerastium vulgatum L. (COMMON MOUSE-EAR CHICKWEED.)
>Occasional at Point Pelee as a weed about dwellings, in gardens and cultivated fields. Islands and Ohio shore.

Cerastium nutans Raf. (NODDING CHICKWEED.)
>Occasional at Point Pelee in sandy shaded ground along "the narrows." Reported near Amherstburg. Islands and Ohio shore.

Agrostemma Githago L. (COMMON COCKLE.)
>Occasional at Point Pelee about dwellings and in waste places, and on Pelee island. Kelley island and Ohio shore.

Lychnis alba Mill. (WHITE CAMPION.)
>Along roads and in old cultivated fields near Windsor.

Silene antirrhina L. (SLEEPY CATCHFLY.)
>Frequent at Point Pelee in sandy open ground and on Pelee island. Kelley island and Ohio shore.

Silene dichotoma Ehrh. (FORKED CATCHFLY.)
>Ohio shore.

Silene noctiflora L. (NIGHT-FLOWERING CATCHFLY.)
>Occasional at Point Pelee about dwellings and in old fields.

Silene virginica L. (FIRE PINK.)
>Put-in-Bay, Kelley, Hartshorn, and Johnson islands and Ohio shore.

Silene latifolia (Mill.) Britten and Rendle. (BLADDER CAMPION.)
>Kelley island.

Saponaria officinalis L. (Bouncing Bet.)

Abundant in spots at Point Pelee in open sandy ground.
Islands and Ohio shore.

Dianthus barbatus L. (Sweet William.)

Frequently about Windsor. (F. P. Cravin.)

PORTULACACEAE (Purslane Family.)

Claytonia virginica L. (Spring Beauty.)

Common at Point Pelee in damp rich woods and thickets.
Along Detroit river. Islands and Ohio shore.

Portulaca oleracea L. (Common Purslane.)]

Common at Point Pelee as a weed in sandy gardens and
field. Islands and Ohio shore.

CERATOPHYLLACEAE (Hornwort Family.)

Ceratophyllum demersum L. (Hornwort.)

In ponds and slow streams near Detroit river. (Maclagan.) Put-in-Bay island.

NYMPHAEACEAE (Water Lily Family.)

Nymphaea advena Ait. (Yellow Water Lily.)

About the ponds and small lakes and in very wet places
of the big marsh at Point Pelee and on Pelee island.
About Middle Bass island.

Nymphaea advena Ait., var. **variegata** (Engelm.) Fernald.
(Variegated Yellow Water Lily.)

In Sandusky bay, Ohio shore.

Castalia tuberosa (Paine) Greene. (White Water Lily.)

Common on the big ditches, ponds, and small lakes of
the big marsh. Rare on Ohio shore.

Nelumbo lutea (Willd.) Pers. (Yellow Nelumbo.)

Abundant in Sandusky bay, Ohio shore.

Brasenia Schreberi Gmel. (Water Shield.)

Frequent at Point Pelee in the ponds and small lakes of
the big marsh. Rare on Ohio shore.

RANUNCULACEAE (Crowfoot Family.)

Ranunculus circinatus Sibth. (Stiff Water Crowfoot.)
Frequent in stagnant water on Pelee island. In wet places along north shore of Lake Erie.

Ranunculus delphinifolius Torr. (Yellow Water Crowfoot.)
Occasional at Point Pelee in ditches and wet places. Islands and Ohio shore.

Ranunculus sceleratus L. (Cursed Crowfoot.)
Frequent at Point Pelee on borders of the big marsh. Islands and Ohio shore.

Ranunculus abortivus L. (Small-flowered Crowfoot.)
Common at Point Pelee in rich woods and thickets. Islands and Ohio shore.

Ranunculus recurvatus Poir. (Hooded Crowfoot.)
Frequent at Point Pelee in open woods and thickets, and on Pelee island.

Ranunculus fascicularis Muhl. (Early Crowfoot.)
Near Detroit river below Amherstburg. Johnson and Kelley islands and Ohio shore.

Ranunculus septentrionalis Poir. (Swamp Buttercup.)
Occasional at Point Pelee in damp open or shaded ground. Put-in-Bay and Kelley islands and Ohio shore.

Ranunculus hispidus Michx. (Hispid Buttercup.)
Reported in wet ground along Detroit river.

Ranunculus repens L. (Creeping Buttercup.)
Frequent about Windsor. (F. P. Cravin.)

Ranunculus pennsylvanicus L.f. (Bristly Crowfoot.)
In damp grassy places along north shore of Lake Erie. Ohio shore.

Ranunculus acris L. (Tall Crowfoot.)
Noticed about Leamington and Windsor. Put-in-Bay is' nd.

Thalictrum dioicum L. (Early Meadow Rue.)
Abundant at Point Pelee in shaded ground along "the narrows" and in rich open woods. Islands and Ohio shore.

Thalictrum dasycarpum Fisch and Lall. (PURPLISH MEADOW RUE.)

 Common at Point Pelee along "the narrows" and in rich open woods. Ohio shore.

Thalictrum polygamum Muhl. (TALL MEADOW RUE.)

 Ohio shore.

Anemonella thalictroides (L.) Spach. (RUE ANEMONE.)

 Noticed by Prof. John Macoun on Pelee island. The other islands and Ohio shore.

Hepatica triloba Chaix. (ROUND-LOBED LIVERLEAF.)

 In dryish woods about Leamington. Islands and Ohio shore.

Hepatica acutiloba DC. (SHARP-LOBED LIVERLEAF.)

 In rich woods about Leamington. Islands and Ohio shore.

Anemone cylindrica Gray. (LONG-FRUITED ANEMONE.)

 Frequent at Point Pelee in dry ground. Ohio shore.

Anemone virginiana L. (TALL ANEMONE.)

 On Point Pelee island and other islands.

Anemone canadensis L. (CANADA ANEMONE.)

 Occasional at Point Pelee in damp open ground. Islands and Ohio shore.

Anemone quinquefolia L. (WOOD ANEMONE.)

 On margins of woods and thickets at Point Pelee. Common about Windsor. Islands.

Clematis virginiana L. (VIRGIN'S BOWER.)

 Frequent on Pelee island. North Bass island.

Caltha palustris L. (MARSH MARIGOLD.)

 Occasional at Point Pelee in damp open or shaded places.

Coptis trifolia (L.) Salisb. (GOLD THREAD.)

 Common at Windsor. (F. P. Cravin.)

Aquilegia canadensis L. (WILD COLUMBINE.)

 Very abundant at Point Pelee along "the narrows." Islands and Ohio shore.

Actaea rubra (Ait.) Willd. (RED BANEBERRY.)

 Cedar point, Ohio shore.

Actaea alba (L.) Mill. (WHITE BANEBERRY.)

 Ohio shore.

Hydrastis canadensis L. (GOLDEN SEAL.)

 Noticed in rich woods along Detroit river.

MAGNOLIACEAE (Magnolia Family.)

Liriodendron tulipifera L. (Tulip Tree.)

Known in the lumber trade as whitewood or yellow poplar. One tree on the farm of Wallace Tilden at Point Pelec. Fine specimens still existing (1911) on the north shore of Lake Erie west to Detroit river. Ohio shore.

ANONACEAE (Custard Apple Family.)

Asimina triloba Dunal. (Common Papaw.)

Reported by Prof. John Macoun as formerly abundant at Point Pelee. Not noticed there in 1910, but fine specimens seen near Colchester in 1911. Formerly on Kelley island.

MENISPERMACEAE (Moonseed Family.)

Menispermum canadensis L. (Moonseed.)

Frequent at Point Pelee in rich woods and thickets. Islands and Ohio shore.

BERBERIDACEAE (Barberry Family.)

Podophyllum peltatum L. (Mandrake.)

Frequent and in spots abundant at Point Pelee in dryish shaded ground.

Jeffersonia diphylla (L.) Pers. (Twinleaf.)

Frequent in woods about Windsor. (F. P. Cravin.) Johnson island.

Caulophyllum thalictroides (L.) Michx. (Papoose Root.)

Frequent at Point Pelee in rich woods and on Pelee island. Johnson island.

LAURACEAE (Laurel Family.)

Sassafras variifolium (Salisb.) Ktze. (Sassafras.)

Common in dry woods at Point Pelee. Near Leamington, in a small grove on Lake Erie shore, trees were noticed sixty feet high and two feet in diameter three feet from the ground. On Ohio shore some of the trees reported to be two and one-half feet in diameter four feet from the ground.

Benzoin aestivale (L.) Nees. (SPICE BUSH.)
Frequent at Point Pelee in damp open woods and thickets, and on Pelee island. Ohio shore.

PAPAVERACEAE (POPPY FAMILY.)

Sanguinaria canadensis L. (BLOODROOT.)
Frequent at Point Pelee in rich woods and thickets. Islands and Ohio shore.

FUMARIACEAE (FUMITORY FAMILY.)

Dicentra Cucullaria (L.) Bernh. (DUTCHMAN'S BREECHES.)
Noticed in sandy ground at Point Pelee among low junipers. Apparently rare. Islands and Ohio shore.

Dicentra canadensis (Goldie) Walp. (SQUIRREL CORN.)
In rich woods about Windsor. (F. P. Cravin.)

Corydalis flavula (Raf.) DC. (PALE CORYDALIS.)
Plentiful at Point Pelee in rich woods. (P. A. Taverner.) Islands and Ohio shore.

Corydalis aurea Willd. (GOLDEN CORYDALIS.)
Ohio shore.

Fumaria officinalis L. (COMMON FUMITORY.)
Kelley island and Ohio shore.

CRUCIFERAE (MUSTARD FAMILY.)

Draba verna L. (WHITLOW GRASS.)
In sandy waste places and on roadsides about Windsor. (F. P. Cravin.)

Draba caroliniana Walt. (CAROLINA WHITLOW GRASS.)
Plentiful at Point Pelee in dry open ground near Albert Gardner's farm, and on Pelee island. Ohio shore.

Alyssum alyssoides L. (YELLOW ALYSSUM.)

Noticed at Point Pelee as a weed in dry open ground near dwellings. Islands and Ohio shore.

Thlaspi arvense L. (FIELD PENNY CRESS.)

A weed about villages and towns.

Ledipium virginicum L. (WILD PEPPERGRASS.)

Common at Point Pelee in gardens and fields. Islands and Ohio shore.

Ledipium apetalum Willd. (APETALOUS PEPPERGRASS.)

Occasional at Point Pelee near dwellings and in gardens, and on Pelee island. Along Detroit river.

Lepidium campestre (L.) R. Br. (FIELD CRESS.)

Occasional at Point Pelee about dwellings and along roads. Put-in-Bay and Kelley island.

Capsella Bursa-pastoris (L.) Medic. (SHEPHERD'S PURSE.)

A common weed at Point Pelee in gardens and cultivated fields. Abundant on the islands and Ohio shore.

Camelina sativa (L.) Crantz. (FALSE FLAX.)

Occasional at Point Pelee as a weed about dwellings.

Cakile edentula (Bigel.) Hook. (AMERICAN SEA ROCKET.)

Common at Point Pelee on the sandy beach, often below the wave line. Islands and Ohio shore.

Brassica arvensis (L.) Ktze. (COMMON MUSTARD.)

A weed at Point Pelee in cultivated fields but apparently infrequent. Abundant on the islands and Ohio shore.

Brassica nigra (L.) Koch. (BLACK MUSTARD.)

Occasional at Point Pelee as a weed about dwellings and in gardens. Islands and Ohio shore.

Alliaria officinalis Andrz. (GARLIC MUSTARD.)

Kelley island.

Sisymbrium officinale (L.) Scop., var. **leiocarpum** DC. (HEDGE MUSTARD.)

Occasional at Point Pelee as a weed about dwellings. Islands and Ohio shore.

Sisymbrium altissimum L. (TUMBLE MUSTARD.)

A weed in towns and along roads.

Sisymbrium canescens Nutt. (TANSY MUSTARD.)

Common at Point Pelee in dry open or shaded ground. Islands and Ohio shore.

Sisymbrium canescens Nutt., var. **brachycarpon** (Richards.) Wats. (TANSY MUSTARD.)

>At Point Pelee. (Canadian Catalogue of Plants.)

Erysimum cheiranthoides L. (WORM-SEED MUSTARD.)

>Occasional at Point Pelee as a weed about dwellings and in old fields. Pelee island.

Radicula Nasturtium-aquaticum (L.) Britten and Rendle. (TRUE WATER CRESS.)

>In ditches and creeks about Windsor. (F. P. Cravin.)

Radicula sylvestris (L.) Druce. (YELLOW CRESS.)

>Along roads near Leamington.

Radicula palustris (L.) Moench. (MARSH CRESS.)

>Occasional at Point Pelee in wet places on borders of the big marsh. Islands and Ohio shore.

Radicula palustris (L.) Moench, var. **hispida** (Desv.) Robinson. (HISPID YELLOW CRESS.)

>Islands and Ohio shore.

Radicula Armoracia (L.) Robinson. (HORSERADISH.)

>Occasional at Point Pelee as an escape. Islands and Ohio shore.

Barbarea vulgaris R. Br. (COMMON WINTER CRESS.)

>Frequent at Point Pelee as a weed in gardens and fields. Green island and Ohio shore.

Barbarea stricta Andrz. (ERECT-FRUITED WINTER CRESS.)

>Green island.

Dentaria diphylla Michx. (TWO-LEAVED TOOTHWORT.)

>In rich shaded ground about Windsor. (F. P. Cravin.)

Dentaria laciniata Muhl. (CUT-LEAVED TOOTHWORT.)

>In rich shaded ground along Detroit river. Islands.

Cardamine bulbosa (Schreb.) BSP. (SPRING CRESS.)

>Occasional at Point Pelee in damp rich woods. Ohio shore.

Cardamine Douglassii (Torr.) Britton. (PURPLE CRESS.)

>In rich woods at Point Pelee. Islands and Ohio shore.

Cardamine pennsylvanica Muhl. (PENNSYLVANIA BITTER CRESS.)

>In wet shaded places at Point Pelee. Islands and Ohio shore.

Arabis lyrata L. (LYRE-LEAVED ROCK CRESS.)
Common at Point Pelee in open dry ground, and on Pelee island. Ohio shore.

Arabis dentata T. and G. (TOOTHED ROCK CRESS.)
Common at Point Pelee in dry open or shaded ground along "the narrows." Johnson, North Bass, and Green islands.

Arabis glabra (L.) Bernh. (TOWER MUSTARD.)
Occasional at Point Pelee in dry open or slightly shaded ground. Johnson island.

Arabis Drummondi Gray. (PURPLE ROCK CRESS.)
Islands and Ohio shore.

Arabis hirsuta (L.) Scop. (HAIRY ROCK CRESS.)
Mouse island and Ohio shore.

Arabis laevigata (Muhl.) Poir. (SMOOTH ROCK CRESS.)
Plentiful at Point Pelee in dry open woods. Islands and Ohio shore.

Arabis canadensis L. (SICKLE-POD.)
Occasional at Point Pelee in dry open woods and on Pelee island. Johnson, Put-in-Bay, and Middle Bass islands and Ohio shore.

CAPPARIDACEAE (CAPER FAMILY.)

Polanisia graveolens Raf. (CLAMMY-WEED.)
Common at Point Pelee on the sandy beach. It has crept into sandy gardens and fields and become a troublesome weed. Islands and Ohio shore.

RESEDACEAE (MIGNONETTE FAMILY.)

Reseda lutea L. (YELLOW CUT-LEAVED MIGNONETTE.)
Kelley island.

DROSERACEAE (SUNDEW FAMILY.)

Drosera rotundifolia L. (ROUND-LEAVED SUNDEW.)
In swampy places about Windsor. (F. P. Cravin.)

Penthorum sedoides L. (DITCH STONECROP.)

Occasional at Point Pelee in open wet places. Islands and Ohio shore.

Sedum acre L. (MOSSY STONECROP.)

Occasional at and near Point Pelee in dry sandy ground. Kelley island and Ohio shore.

Sedum ternatum Michx. (WILD STONECROP.)

Put-in-Bay island.

Sedum purpureum Tausch. (LIVE-FOR-EYER.)

Occasional at Point Pelee near dwellings as an escape, and on Pelee island. Put-in-Bay and North Bass islands and Ohio shore.

SAXIFRAGACEAE (SAXIFRAGE FAMILY.)

Tiarella cordifolia L. (FALSE MITERWORT.)

In rich woods at Point Pelee. (Wallace Tilden.) Apparently rare.

Heuchera americana L. (COMMON ALUM ROOT.)

On Pelee island and near Amherstburg. The other islands.

Mitella diphylla L. (TWO-LEAVED BISHOP'S CAP.)

Common in rich shaded ground about Windsor.

Parnassia caroliniana Michx. (CAROLINE GRASS OF PARNASSUS.)

In damp spots along north shore of Lake Erie, and near Sandwich.

Ribes Cynosbati L. (PRICKLY GOOSEBERRY.)

Frequent at Point Pelee in damp open woods and thickets. Islands and Ohio shore.

Ribes floridum L'Her. (WILD BLACK CURRANT.)

Common at Point Pelee in damp woods and thickets and on Pelee island. Kelley island and Ohio shore.

Ribes aureum Pursh. (MISSOURI CURRANT.)

Noticed as an occasional escape on Pelee island. Kelley island.

HAMAMELIDACEAE (Witch-Hazel Family.)

Hamamelis virginiana L. (Witch-hazel.)
　　Common at Point Pelee in open dry woods.

PLATANACEAE (Plane Tree Family.)

Platanus occidentalis L. (Sycamore.)
　　Frequent at Point Pelee in rich ground with other trees.
　　Islands and Ohio shore.

ROSACEAE (Rose Family.)

Physocarpus opulifolius (L.) Maxim. (Nine-bark.)
　　Common on the islands and Ohio shore.

Spiraea salicifolia L. (Meadow-sweet.)
　　Frequent at Point Pelee in damp open places.

Pyrus communis L. (Common Pear.)
　　Several large trees, apparent escapes, along north shore
　　　　of Lake Erie. Put-in-Bay and Kelly islands and Ohio
　　　　shore.

Pyrus coronaria L. (American Crab.)
　　Not noticed at Point Pelee, but plentiful farther west
　　along north shore of Lake Erie. Near Amherstburg.
　　(Macoun.) Put-in-Bay island.

Pyrus Malus L. (Apple.)
　　Frequent at Point Pelee in open ground. Islands and
　　　　Ohio shore.

Pyrus cydonia L. (Common Quince.)
　　Several fine looking trees in waste places along north
　　　　shore of Lake Erie, but perhaps not escapes.

Pyrus melanocarpa (Michx.) Willd. (Black Chokeberry.)
　　Common about Windsor. Ohio shore.

Pyrus americana (Marsh.) DC. (American Mountain
Ash.)
　　Occasionally planted in Essex county. Rattlesnake and
　　　　Put-in-Bay islands. Probably from seeds dropped
　　　　by birds.

Amelanchier canadensis (L.) Medic. (COMMON JUNEBERRY.)
> Frequent at Point Pelee in open woods. Islands and Ohio shore.

Amelanchier oblongifolia (T. and G.) Roem. (SHAD-BUSH.)
> Occasional at Point Pelee in open sandy ground. Mouse and Kelley island and Ohio shore.

Amelanchier spicata (Lam.) C. Koch. (ROUND-LEAVED JUNEBERRY.)
> Abundant in spots at Point Pelee in open sandy ground.

Crataegus Oxyacantha L. (ENGLISH HAWTHORNE.)
> Occasionally planted as an ornamental tree and in hedges. but perhaps not yet escaping.

Crataegus Crus-galli L. (COCKSPUR THORN.)
> Occasional at Point Pelee in open places and borders of woods, and on Pelee island. Ohio shore.

Crataegus punctata Jacq. (LARGE-FRUITED THORN.)
> Common at Point Pelee in open places and open woods. One tree noticed near Kingsville, two feet in diameter two feet from ground.

Crataegus tomentosa L. (PEAR THORN.)
> Kelley and Middle Bass islands.

Crataegus mollis (T. and G.) Schelle. (RED-FRUITED THORN.)
> Frequent at Point Pelee and on Pelee island. Along Detroit river. Kelley island and Ohio shore.

Fragaria virginiana Duchesne. (COMMON STRAWBERRY.)
> Common at Point Pelee and on Pelee island. Kelley, Put-in-Bay, and Mouse islands and Ohio shore.

Fragaria virginiana Duchesne, var. **illinoensis** (Prince) Gray. (ILLINOIS STRAWBERRY.)
> Kelley, Put-in-Bay, and Mouse islands and Ohio shore.

Fragaria vesca L., var. **americana** Porter. (AMERICAN WOOD STRAWBERRY.)
> Frequent at Point Pelee in rich open woods. Kelley and Put-in-Bay islands and Ohio shore.

Waldsteinia fragarioides (Michx.) Trattinick. (BARREN STRAWBERRY.)
> Frequent about Windsor. (F. P. Cravin.)

Potentilla arguta Pursh. (TALL CINQUEFOIL.)
>Put-in-Bay island and Ohio shore.

Potentilla monspeliensis L. (ROUGH CINQUEFOIL.)
>Occasional at Point Pelee as a weed about dwellings and
in old fields, and on Pelee island. Put-in-Bay island
and Ohio shore.

Potentilla paradoxa Nutt. (BUSHY CINQUEFOIL.)
>Abundant in spots at Point Pelee in open sandy ground
and along north shore of Lake Erie.

Potentilla argentea L. (SILVERY CINQUEFOIL.)
>Common about Windsor on poor open ground.

Potentilla palustris (L.) Scop. (MARSH FIVE-FINGER.)
>Frequent at Point Pelee in wet places in and about the
big marsh.

Potentilla Anserina L. (SILVER WEED.)
>Common at Point Pelee and often abundant, usually
in dry sandy ground near the lake shore. Pelee island,
Middle Bass, North Bass, and Rattlesnake islands,
and Ohio shore.

Potentilla canadensis L. (COMMON CINQUEFOIL.)
>Occasional at Point Pelee in dry open ground, and on
Pelee island. Ohio shore.

Potentilla canadensis L., var. **simplex** (Michx.) T. and G.
(LARGER COMMON CINQUEFOIL.)
>Near Amherstburg. (Macoun.)

Filipendula rubra (Hill) Robinson. (QUEEN OF THE PRAIRIE.)
>Ohio shore.

Geum canadense Jacq. (WHITE AVENS.)
>Common at Point Pelee in rich woods and thickets.
Common on the islands and Ohio shore.

Geum virginianum L. (ROUGH AVENS.)
>Pelee island. Along Detroit river. Put-in-Bay island.

Geum strictum Ait. (YELLOW AVENS.)
>Common about Windsor. (F. P. Cravin.)

Geum vernum (Raf.) T. and G. (SPRING AVENS.)
>Pelee island. Near Amherstburg. (Macoun.) Common
about Windsor. Johnson island and Ohio shore.

Geum rivale L. (WATER AVENS.)

> In damp open or shaded ground about Windsor. (F. P. Cravin.)

Rubus idaeus L., var. **canadensis** Richardson (*R. strigosus* Michx. of most authors). (WILD RED STRAWBERRY.) See Rhodora XI–236.

> Occasional at Point Pelee in dry open places, and on Pelee island. Ohio shore.

Rubus neglectus Peck. (PURPLE WILD RASPBERRY.)

> Near Amherstburg. (Macoun.)

Rubus occidentalis L. (BLACK RASPBERRY.)

> Frequent at Point Pelee in rich open woods. Islands and Ohio shore.

Rubus odoratus L. (PURPLE FLOWERING RASPBERRY.)

> A large bunch on the bank of a big ditch in the big marsh at Point Pelee. Not noticed elsewhere.

Rubus triflorus Richards. (DWARF RASPBERRY.)

> Common in damp shaded ground about Windsor.

Rubus alleghaniensis Porter. (*R. villosus* of many authors.) (HIGH BUSH BLACKBERRY.)

> Frequent at Point Pelee in dry open and often shaded ground. Islands and Ohio shore.

Rubus hispidus L. (RUNNING SWAMP BLACKBERRY.)

> Common about Windsor.

Rubus villosus Ait. (*R. canadensis* of most authors.) (DEWBERRY.)

> Occasional at Point Pelee in dry sandy ground. Islands and Ohio shore.

Agrimonia gryposepala Wallr. (TALL HAIRY AGRIMONY.)

> Frequent at Point Pelee in open or slightly shaded ground especially along "the narrows," and on Pelee island. Kelley island and Ohio shore.

Agrimonia parviflora Ait. (MANY-FLOWERED AGRIMONY.)

> Frequent at Point Pelee in dry open ground, and on Pelee island. Along north shore of Lake Erie. Ohio shore.

Rosa setigera Michx. (CLIMBING ROSE.)

Pelee island, and near Amherstburg. (Macoun.) South shore of Lake St. Clair. Johnson, Mouse, Kelley, and Middle Bass islands, and Ohio shore.

Rosa blanda Ait. (MEADOW ROSE.)

Pelee island, Old Hen island, and Ohio shore.

Rosa rubiginosa L. (SWEETBRIER.)

Common at Point Pelee on flat, dry, open ground. Islands and Ohio shore.

Rosa carolina L. (SWAMP ROSE.)

Frequent at Point Pelee in low damp open places. Islands and Ohio shore.

Rosa humilis Marsh. (PASTURE ROSE.)

Along Detroit river and near Leamington. (Macoun.) Kelley and Put-in-Bay islands and Ohio shore.

Prunus serotina Ehrh. (WILD BLACK CHERRY.)

Frequent at Point Pelee, especially along "the narrows." Islands and Ohio shore.

Prunus virginiana L. (CHOKE CHERRY.)

Frequent at Point Pelee on border of woods. Islands and Ohio shore.

Prunus pennsylvanica L. f. (WILD RED CHERRY.)

Common along north shore of Lake Erie.

Prunus institita L. (BULLACE PLUM.)

Noticed growing wild on Pelee island by Prof. John Macoun.

Prunus pumila L. (SAND CHERRY.)

Frequent at Point Pelee in sandy open ground and on upper beach. Ohio shore.

Prunus avium L. (SWEET CHERRY.)

A few trees noticed along shore of Lake Erie, apparently permanent escapes. Kelley island.

Prunus Cerasus L. (COMMON CHERRY.)

Frequent as an escape at Point Pelee and on Pelee island.

Prunus americana Marsh. (WILD PLUM.)

Noticed on Pelee island. Along north shore of Lake Erie. Kelley and Put-in-Bay islands.

Prunus Persica (L.) Stokes. (PEACH.)

Inclined to persist on the islands of Lake Erie, and Ohio shore.

LEGUMINOSAE (PULSE FAMILY.)

Gymnocladus dioica (L.) Koch. (KENTUCKY COFFEE-TREE.)

Noticed in 1882 on Pelee island by Prof. John Macoun. Not seen in 1912. The other islands and Ohio shore.

Gleditsia triacanthos L. (HONEY LOCUST.)

Many trees large and small growing on and near the beach on the west side of Point Pelee, and occasional along north shore of Lake Erie to Detroit river. The islands and Ohio shore.

Cassia marilandica L. (WILD SENNA.)

Johnson island and Ohio shore.

Cassia Chamaecrista L. (PARTRIDGE PEA.)

Ohio shore.

Cercis canadensis L. (REDBUD.)

Pelee island. (Macoun.) Ohio shore.

Baptisia tinctoria (L.) R. Br. (WILD INDIGO.)

Near Colchester and Sandwich. (Macoun.) Near Leamington and Windsor.

Lupinus perennis L. (WILD LUPINE.)

Near Sandwich and Leamington in dry open ground.

Trifolium arvense L. (RABBIT-FOOT CLOVER.)

In waste places about Windsor. (F. P. Cravin.)

Trifolium pratense L. (RED CLOVER.)

Frequent at Point Pelee in open ground. Islands and Ohio shore.

Trifolium reflexum L. (BUFFALO CLOVER.)

Islands of Detroit river. Johnson island.

Trifolium repens L. (WHITE CLOVER.)

Occasional at Point Pelee in old fields and pastures. Islands and Ohio shore.

Trifolium hybridum L. (ALSIKE CLOVER.)

Occasional at Point Pelee on and near cultivated ground. Put-in-Bay island and Ohio shore.

Trifolium agrarium L. (YELLOW CLOVER.)
> Frequent about Windsor. (F. P. Cravin.)

Trifolium procumbens L. (LOW HOP CLOVER.)
> Gibraltar island in Lake Erie.

Melilotus officinalis (L.) Lam. (YELLOW MELILOT.)
> Occasional at Point Pelee near dwellings. Johnson and Put-in-Bay islands and Ohio shore.

Melilotus alba Desr. (SWEET CLOVER.)
> Occasional at Point Pelee about dwellings and road-sides. Common along Detroit river. Abundant in fields about Windsor. Abundant on the islands and Ohio shore.

Medicago sativa L. (ALFALFA.)
> Rare at Point Pelee. Occasional throughout Essex county. Put-in-Bay island and Ohio shore.

Medicago lupulina L. (BLACK MEDICK.)
> Frequent at Point Pelee about dwellings. Islands and Ohio shore.

Robinia Pseudo-Acacia L. (COMMON LOCUST.)
> Inclined to escape and persist at Point Pelee. Islands and Ohio shore.

Astragalus canadensis L. (CAROLINA MILK VETCH.)
> Frequent along the north shore of Lake Erie. Islands and Ohio shore.

Desmodium nudiflorum (L.) DC. (NAKED-FLOWERED TICK TREFOIL.)
> Ohio shore.

Desmodium grandiflorum (Walt.) DC. (POINTED-LEAVED TICK TREFOIL.)
> Ohio shore.

Desmodium rotundifolium (Michx.) DC. (POSTRATE TICK TREFOIL.)
> Ohio shore.

Desmodium canescens (L.) DC. (HOARY TICK TREFOIL.)
> Occasional at Point Pelee in sandy open ground, and along north shore of Lake Erie. Islands and Ohio shore.

Desmodium bracteosum (Michx.) DC. (LARGE-BRACTED
TICK TREFOIL.)
>Along north shore of Lake Erie. Ohio shore.

Desmodium illinoense Gray. (ILLINOIS TICK TREFOIL.)
>Ohio shore.

Desmodium Dillenii Darl. (DILLEN'S TICK TREFOIL.)
>Occasional at Point Pelee in dry shaded ground, and along
>north shore of Lake Erie. Put-in-Bay island.

Desmodium paniculatum (L.) DC. (PANICLED TICK TRE-
FOIL.)
>Occasional at Point Pelee in dry open or slightly shaded
>ground. Put-in-Bay island and Ohio shore.

Desmodium canadense (L.) DC. (SHOWY TICK TREFOIL.)
>North shore of Lake Erie and along Detroit river. Along
>Lake St. Clair. Ohio shore.

Desmodium sessilifolium (Torr.) T. and G. (SESSILE-LEAVED
TICK TREFOIL.)
>Dry shaded ground near Amherstburg.

Desmodium rigidum (Ell.) DC. (RIGID TICK TREFOIL.)
>Ohio shore.

Lespedeza violacea (L.) Pers. (BUSH CLOVER.)
>Ohio shore.

Lespedeza virginica (L.) Britton.
>In thickets along Detroit river and near Leamington.
>(Macoun.)

Lespedeza frutescens (L.) Britton. (WAND-LIKE BUSH CLO-
VER.)
>Sandy woods and thickets in Essex county. (Macoun.)
>Ohio shore.

Lespedeza hirta (L.) Hornem. (HAIRY BUSH CLOVER.)
>Ohio shore.

Lespedeza capitata Michx. (ROUND-LEAVED BUSH CLOVER.)
>Along Detroit river. Ohio shore.

Vicia sativa L. (SPRING VETCH.)
>North Bass and Rattlesnake islands and Ohio shore.

Vicia angustifolia (L.) Reichard. (COMMON VETCH.)
>Frequent as a weed about Windsor.

Vicia caroliniana Walt. (CAROLINA VETCH.)
    Along Detoit river. Islands and Ohio shore.
Vicia americana Muhl. (AMERICAN VETCH.)
    Kelley and North Bass islands and Ohio shore.
Lathyrus maritimus (L.) Bigel. (BEACH PEA.)
    Along north shore of Lake Erie. Apparently infrequent.
    Ohio shore.
Lathyrus palustris L. (MARSH VETCHLING.)
    Frequent at Point Pelee on damp meadow-like ground.
    Islands and Ohio shore.
Lathyrus ochroleucus Hook. (CREAM-COLOURED VETCH-
    LING.)
    Islands and Ohio shore.
Apios tuberosa Moench. (WILD BEAN.)
    Common at Point Pelee along "the narrows" and borders
    of woods and thickets, and on Pelee island. Ohio shore.
Strophostyles helvola (L.) Britton. (TRAILING WILD BEAN.)
    Frequent and often abundant at Point Pelee in damp
    sand, and along the beach. Islands and Ohio shore.
Amphicarpa monoica (L.) Ell. (HOG PEANUT.)
    Frequent at Point Pelee in dry open ground and open
    woods. Islands and Ohio shore.
Amphicarpa Pitcheri T. and G. (PITCHER's HOG PEANUT.)
    Frequent at Point Pelee along "the narrows." Islands
    and Ohio shore.

LINACEAE (FLAX FAMILY.)

Linum usitatissimum L. (COMMON FLAX.)
    Occasional at Point Pelee about farm buildings. Kelley
    island.
Linum sulcatum Riddell. (GROOVED YELLOW FLAX.)
    Occasional about Windsor. (F. P. Cravin.) Ohio shore.

OXALIDACEAE (WOOD SORREL FAMILY.)

Oxalis violacea L. (VIOLET WOOD SORREL.)
    Ohio shore.

Oxalis corniculata L. (LADY'S SORREL.)

Frequent at Point Pelee in meadow-like ground, gardens, and fields. Islands and Ohio shore.

GERANIACEAE (GERANIUM FAMILY.)

Geranium maculatum L. (WILD CRANESBILL.)

Common at Point Pelee in open ground and open woods, and on Pelee island. Kelley island and Ohio shore.

Geranium Robertianum L. (HERB ROBERT.)

Abundant at Point Pelee under red cedars and pines or in damp rich woods. Islands and Ohio shore.

Geranium carolinianum L. (CAROLINA CRANESBILL.)

Occasional at Point Pelee in dry open ground. Islands and Ohio shore.

RUTACEAE (RUE FAMILY.)

Zanthoxylum americanum Mill. (NORTHERN PRICKLY ASH.)

Common at Point Pelee in and about borders of woods and thickets, and on Pelee island. Kelley and Middle Bass islands and Ohio shore.

Ptelea trifoliata L. (SHRUBBY TREFOIL.)

Frequent at Point Pelee along "the narrows" and in other places. Reported as formerly much more abundant. Known at Point Pelee as "Wahoo." Along north shore of Lake Erie. Plentiful on Pelee island. Common on the other islands and Ohio shore.

SIMARUBACEAE (QUASSIA FAMILY.)

Ailanthus glandulosa Desf. (TREE-OF-HEAVEN.)

Escaping in Essex county. Islands and Ohio shore.

POLYGALACEAE (MILKWORT FAMILY.)

Polygala polygama Walt. (RACEMED MILKWORT.)

In dry ground about Windsor (F. P. Cravin.)

Polygala Senega L (SENECA SNAKEROOT.)
Ohio shore.

Polygala Senega L., var. **latifolia** T. and G. (SENECA SNAKEROOT.)
Ohio shore.

Polygala sanguinea L. (PURPLE MILKWORT.)
Point Pelee. (Burgess.) Near Windsor (J. M. Macoun.)

Polygala verticillata L. (WHORLED MILKWORT.)
In dry open ground from Amherstburg to Sandwich. Ohio shore.

EUPHORBIACEAE (SPURGE FAMILY.)

Acalypha virginica L. (VIRGINIA THREE-SEEDED MERCURY.)
Frequent at Point Pelee in open places and often abundant as a weed in cultivated fields. Islands and Ohio shore.

Euphorbia polygonifolia L. (SEASIDE SPURGE.)
A characteristic plant of the sandy beach at Point Pelee. Islands and Ohio shore.

Euphorbia serpens HBK. (ROUND-LEAVED SPREADING SPURGE.)
Johnson island.

Euphorbia Preslii Guss. (UPRIGHT SPOTTED SPURGE.)
Occasional at Point Pelee in dry open ground. Islands and Ohio shore.

Euphorbia hirsuta (Torr.) Wiegand. (HAIRY SPURGE.)
Along railway embankments about Windsor.

Euphorbia maculata L. (MILK PURSLANE.)
Frequent at Point Pelee near dwellings and on roadsides. Islands and Ohio shore.

Euphorbia corollata L. (FLOWERING SPURGE.)
Occasional at Point Pelee in open dry ground. Ohio shore.

Euphorbia dentata Michx. (TOOTHED SPURGE.)
Islands and Ohio shore.

Euphorbia platyphylla L. (BROAD-LEAVED SPURGE.)
Near Colchester and along Detroit river.

Euphorbia Helioscopia L. (WARTWEED.)
Near Essex Centre. (Burgess.)

Euphorbia Cyparissias L. (CYPRESS SPURGE.)
Occasional at Point Pelee in open dry ground. Islands and Ohio shore.

Euphorbia commutata Engelm. (TINTED SPURGE.)
Johnson island.

LIMNANTHACEAE (FALSE MERMAID FAMILY.)

Floerkea proserpinacoides Willd. (FALSE MERMAID.)
On an island of Detroit river near Amherstburg.

ANACARDIACEAE (CASHEW FAMILY.)

Rhus typhina L. (STAGHORN SUMACH.)
Frequent at Point Pelee in dry open or slightly shaded ground. Islands and Ohio shore.

Rhus glabra L. (SMOOTH SUMACH.)
Plentiful on Pelee island. The other islands and Ohio shore.

Rhus Vernix L. (POISON SUMACH.)
Frequent at Point Pelee in damp ground along "the narrows" and in swampy places, and on Pelee island.

Rhus Toxicodendron L. (POISON IVY.)
Very common at Point Pelee in dry or damp, rich, open or shaded ground. Often taking complete possession of trees forty and fifty feet high, the vines being in many cases more than three inches in diameter one foot from the ground. A low shrub when not climbing. The climbing form often referred to a s*R. Toxicodendron* L., var. *radicans* (L.) Torr., climbing poison ivy. Islands and Ohio shore.

Rhus canadensis Marsh. (FRAGRANT SUMACH.)

Very abundant at Point Pelee in dry woods and open dry ground. Also abundant in sand on the west beach where, with low junipers in large and thick bunches, it forms an efficient sand binder against the action of wind. Islands and Ohio shore.

AQUIFOLIACEAE (HOLLY FAMILY.)

Ilex verticillata (L.) Gray. (BLACK ALDER.)

Occasional at Point Pelee in damp thickets, and on Pelee island. Green island.

Ilex verticillata (L.) Gray, var. **tenuifolia** (Torr.) Wats. (BRONX WINTERBERRY.)

Near Sandwich and Leamington. (Macoun.)

CELASTRACEAE (STAFF TREE FAMILY.)

Evonymus atropurpureus Jacq. (BURNING BUSH.)

Occasional at Point Pelee in shaded ground. (Wallace Tilden.) Pelee island. White island in Detroit river, Hen and Kelley islands, and Ohio shore.

Evonymus obovatus Nutt. (RUNNING STRAWBERRY BUSH.)

Frequent at Point Pelee in rich shaded ground. Along Lake St. Clair. Islands and Ohio shore.

Celastrus scandens L. (BITTERSWEET.)

Very common at Point Pelee in woods and thickets. Islands and Ohio shore.

STAPHYLEACEAE (BLADDER NUT FAMILY.)

Staphylea trifolia L. (AMERICAN BLADDER NUT.)

Occasional near Windsor. (F. P. Cravin.) Green island and Ohio shore.

ACERACEAE (MAPLE FAMILY.)

Acer spicatum Lam. (MOUNTAIN MAPLE.)

In rich woods about Windsor. (F. P. Cravin.)

Acer saccharum Marsh. (*A. saccharinum* Wang.) (SUGAR MAPLE.)

> Occasional at Point Pelee with other trees. Islands and Ohio shore.

Acer saccharum Marsh, var. **nigrum** (Michx. f.) Britton. (BLACK MAPLE.)

> Along "the narrows" at Point Pelee, and on Pelee island. Kelley and North Bass islands, and Ohio shore. A southern fern was noticed on Pelee island, named by Prof. C. S. Sargent, *A. saccharum rugelii* Rehd. Leaves three-lobed.

Acer saccharinum L. (*A. dasycarpum* Ehrh.) (SILVER MAPLE.)

> Occasional at Point Pelee in rich ground with other trees. Islands and Ohio shore.

Acer rubrum L. (RED MAPLE.)

> Frequent at Point Pelee in rich ground with other trees, and on Pelee island.

Acer Negundo L. (BOX ELDER.)

> Occasionally planted and escaping at Point Pelee and on Pelee island. Put-in-Bay island and Ohio shore.

SAPINDACEAE (SOAPBERRY FAMILY.)

Aesculus Hippocastanum L. (COMMON HORSE-CHESTNUT.)

> Often planted in Essex county, and apparently spreading along north shore of Lake Erie.

Aesculus glabra Willd. (FETID BUCKEYE.)

> Johnson, Middle Bass, and North Bass islands and Ohio shore.

BALSAMINACEAE (TOUCH-ME-NOT FAMILY.)

Impatiens pallida Nutt. (PALE TOUCH-ME-NOT.)

> Frequent about Windsor. (F. P. Cravin.) Old Hen and Rattlesnake islands.

Impatiens biflora Walt. (SPOTTED TOUCH-ME-NOT.)

> Common at Point Pelee in wet open or shaded places. Islands and Ohio shore.

RHAMNACEAE (Buckthorn Family.)

Rhamnus alnifolia L'Her. (Alder-leaved Buckthorn.)
   In damp open ground near Lake St. Clair.
Ceanothus americanus L. (New Jersey Tea.)
   Frequent at Point Pelee in dry open or shaded ground.
   Ohio shore.
Ceanothus ovatus Desf. (Smaller Red-root.)
   Occasional at Point Pelee in dry open or shaded ground.
   Ohio shore.

VITACEAE (Vine Family.)

Psedera vitacea (Knerr) Greene. (*Ampelopsis quinquefolia*
   of most authors). (Virginia Creeper.)
   Common at Point Pelee in shaded ground. Islands
   and Ohio shore.
Vitis labrusca L. (Northern Fox Grape.)
   Many patches of the cultivated grape persisting at Point
   Pelee and on Pelee island apparently belong to or
   are varieties of this species.
Vitis bicolor Le Conte. (Summer Grape.)
   Frequent at Point Pelee in shaded or open ground and
   on Pelee island.
Vitis cordifolia Michx. (Frost Grape.)
   Johnson island.
Vitis vulpina L. (River-bank Grape.)
   Common at Point Pelee in open or shaded ground.
   Islands and Ohio shore.

TILIACEAE (Linden Family.)

Tilia americana L. (Basswood.)
   Common at Point Pelee in rich ground with other trees.
   Islands and Ohio shore.
Tilia Michauxii Nutt. (Southern Basswood.)
   Reported as growing near Sandwich and along Lake
   St. Clair.

MALVACEAE (Mallow Family.)

Abutilon Theophrasti Medic. (Indian Mallow.)
Occasional at Point Pelee as a garden weed. Common on the islands.

Althaea rosea Cav. (Hollyhock.)
Inclined to escape and persist on the islands in Lake Erie. Ohio shore.

Sida spinosa L. (Prickly Sida.)
Kelley island and Ohio shore.

Malva rotundifolia L. (Common Mallow.)
Frequent at Point Pelee as a weed about dwellings and waste places. Islands and Ohio shore.

Malva moschata L. (Musk Mallow.)
Along north shore of Lake Erie and on islands in Detroit river. Kelley island.

Hibiscus Moscheutos L. (Swamp Rose Mallow.)
Occasional at Point Pelee on borders of the big marsh. Reported as formerly abundant. Abundant in many places along the north shore of Lake Erie. Islands of Detroit river. Along shore of Lake St. Clair. North Bass island. Dr. E. L. Greene has named this plant *H. opulifolius*, believing it to be a new species.

Hibiscus Trionum L. (Flower-of-an-hour.)
Kelley island and Ohio shore.

HYPERICACEAE (St. John's-wort Family.)

Hypericum perforatum L. (Common St. John's-wort.)
Frequent at Point Pelee along roads and in old fields and on Pelee island, Kelley island, and Ohio shore.

Hypericum punctatum Lam. (Spotted St. John's-wort.)
Occasional at Point Pelee in shaded ground. Rattlesnake island and Ohio shore.

Hypericum prolificum L. (Shrubby St. John's-wort.)
Frequent about Windsor. (F. P. Cravia.)

Hypericum Kalmianum L. (Kalm's St. John's-wort.)
Islands in Lake Erie.

Hypericum mutilum L. (DWARF ST. JOHN'S-WORT.)

Frequent in low open ground about Windsor. (F. P. Cravin.) Ohio shore.

Hypericum majus (Gray) Britton. (LARGER CANADIAN ST. JOHN'S-WORT.)

In damp open ground about the big marsh at Point Pelee.

Hypericum gentianoides (L.) BSP. (ORANGE GRASS.)

In sandy fields at Sandwich. (Macoun.)

Hypericum virginicum L. (MARSH ST. JOHN'S-WORT.)

Common at Point Pelee in and about the big marsh. Ohio shore.

CISTACEAE (ROCKROSE FAMILY.)

Helianthemum canadense (L.?) Michx. (LONG-BRANCHED FROSTWEED.)

Near Sandwich. (Macoun.) In dry open ground near Lake St. Clair.

Helianthemum majus BSP. (HAIRY FROSTWEED.)

On Cedar point, Ohio shore.

Lechea villosa Ell. (LARGE PINWEED.)

Very common about Windsor in dry open ground. Cedar point, Ohio shore.

Lechea minor L. (THYME-LEAVED PINWEED.)

In open or slightly shaded ground near Windsor.

VIOLACEAE (VIOLET FAMILY.)

Viola cucullata Ait. (MARSH BLUE VIOLET.)

Common about Windsor in swampy places.

Viola palmata L. (EARLY BLUE VIOLET.)

In damp woods near Amherstburg. Ohio shore.

Viola sororia Willd. (WOOLLY BLUE VIOLET.)

Common at Point Pelee in shaded ground. Ohio shore.

Viola fimbriatula Sm. (OVATE-LEAVED VIOLET.)

Common about Windsor.

Viola sagittata Ait. (ARROW-LEAVED VIOLET.)
Near Amherstburg. Common about Windsor.

Viola pedatifida G. Don. (PRAIRIE VIOLET.)
Ohio shore, but scarce.

Viola
Frequent at Point Pelee in rich shaded ground. Islands
and Ohio shore.

Viola pubescens Ait. (DOWNY YELLOW VIOLET.)
Frequent at Point Pelee in dry shaded ground. Islands
and Ohio shore.

Viola scabriuscula Schwein. (SMOOTH YELLOW VIOLET.)
Frequent at Point Pelee in rich woods and thickets.
Islands and Ohio shore.

Viola striata Ait. (PALE VIOLET.)
Near Amherstburg. Ohio shore.

Viola Rafinesquii Greene. (WILD PANSY.)
On Pelee island. (Macoun.) Johnson and Put-in-Bay
islands and Ohio shore.

CACTACEAE (CACTUS FAMILY.)

Opuntia Rafinesquii Engelm. (WESTERN PRICKLY PEAR.)
Abundant in spots at Point Pelee in dry open ground.

THYMELAEACEAE (MEZEREUM FAMILY.)

Dirca palustris L. (LEATHERWOOD.)
Frequent in shaded ground near Colchester.

ELAEAGNACEAE (OLEASTER FAMILY.)

Shepherdia canadensis (L.) Nutt. (BUFFALO BERRY.)
Common at Point Pelee in dry open or shaded ground.

LYTHRACEAE (LOOSESTRIFE FAMILY.)

Rotala ramosior (L.) Koehne. (ROTALA.)
Ohio shore, but rare.

Ammannia coccinea Rottb. (LONG-LEAVED AMMANIA.)

>   Ohio shore.

Decodon verticillatus (L.) Ell. (WATER WILLOW.)

>   Very abundant in spots at Point Pelee in and about the
>   big marsh. Islands and Ohio shore.

Lythrum alatum Pursh. (WING-ANGLED LOOSESTRIFE.)

>   In damp places along north shore of Lake Erie, Detroit
>   river, and east of Windsor. Put-in-Bay and Middle
>   Bass islands and Ohio shore.

ONAGRACEAE (EVENING PRIMROSE FAMILY.)

Ludvigia alternifolia L. (SEEDBOX.)

>   In swampy ground, meadows, and pastures near Sand-
>   wich and Windsor. (Alex. Wherry.)

Ludvigia polycarpa Short and Peter. (MANY-FRUITED LUD-
VIGIA.)

>   In damp ground along railway track near Amherstburg.
>   (Macoun.)

Ludvigia palustris (L.) Ell. (WATER PURSLANE.)

>   In wet places at Point Pelee. Ohio shore.

Epilobium angustifolium L. (GREAT WILLOW-HERB.)

>   Ohio shore.

Epilobium densum Raf. (LINEAR-LEAVED WILLOW-HERB.)

>   Ohio shore.

Epilobium coloratum Muhl. (PURPLE-LEAVED WILLOW-
HERB.)

>   Kelley and Middle Bass islands and Ohio shore.

Epilobium adenocaulon Hausek. (NORTHERN WILLOW-HERB.)

>   Occasional at Point Pelee in damp open ground and on
>   Pelee island. Kelley and North Bass islands and
>   Ohio shore.

Oenothera biennis L. (COMMON EVENING PRIMROSE.)

>   Common at Point Pelee in dry open or shaded ground,
>   and often on the sandy beach. Islands and Ohio shore.

Oenothera rhombipetala Nutt. (RHOMBIC EVENING PRIM-
ROSE.)

>   Ohio shore.

Gaura biennis L. (BIENNIAL GAURA.)

Along Detroit river and near Windsor. Ohio shore.

Circaea lutetiana L. (ENCHANTER'S NIGHTSHADE.)

Common at Point Pelee in rich open woods, and on Pelee island. Put-in-Bay island.

Circaea alpina L. (SMALLER ENCHANTER'S NIGHTSHADE.)

Ohio shore.

HALORAGIDACEAE (WATER MILFOIL FAMILY.)

Myriophyllum spicatum L. (SPIKED WATER MILFOIL.)

Often abundant at Point Pelee in big ditches and stagnant pools, and on Pelee island.

Proserpinaca palustris L. (MERMAID-WEED.)

Common at Point Pelee in and about the big marsh. Ohio shore.

ARALIACEAE (GINSENG FAMILY.)

Aralia racemosa L. (SPIKENARD.)

In rich woods near Colchester.

Aralia nudicaulis L. (WILD SARSAPARILLA.)

Common at Point Pelee in rich moist woods and thickets. Green and Kelley islands and Ohio shore.

Panax quinquefolium L. (GINSENG.)

Reported as formerly plentiful at Point Pelee and on Pelee island, but not noticed in 1910-11. Formerly abundant on the other islands and Ohio shore.

Panax trifolium L. (GROUND-NUT.)

In rich woods about Windsor. (F. P. Cravin.)

UMBELLIFERAE (PARSLEY FAMILY.)

Sanicula marilandica L. (BLACK SNAKEROOT.)

Common at Point Pelee and on Pelee island. Put-in-Bay island and Ohio shore.

Sanicula canadensis L. (SHORT-STYLED SNAKEROOT.)

Along Detroit river. Kelley island and Ohio shore.

Erigenia bulbosa (Michx.) Nutt. (HARBINGER-OF-SPRING.)
Kelley island and Ohio shore.

Chaerophyllum procumbens (L.) Crantz. (SPREADING CHERVIL.)
White island in Detroit river. (Macoun.) Kelley island.

Osmorhiza Claytoni (Michx.) Clarke. (WOOLLY SWEET CICELY.)
Islands and Ohio shore.

Osmorhiza longistylis (Torr.) DC. (SMOOTHER SWEET CICELY.)
Plentiful at Point Pelee in rich open or shaded ground. Islands and Ohio shore.

Conium maculatum L. (POISON HEMLOCK.)
Noticed in waste places near Windsor.

Cicuta maculata L. (WATER HEMLOCK.)
In damp open ground on Pelee island. Kelley island and Ohio shore.

Cicuta bulbifera L. (BULB-BEARING WATER HEMLOCK.)
Common at Point Pelee in and about the big marsh. Islands and Ohio shore.

Carum Carvi L. (CARAWAY.)
Occasional at Point Pelee as a weed about dwellings. Islands and Ohio shore.

Sium cicutaefolium Schrank. (WATER PARSNIP.)
Common at Point Pelee in and about the big marsh and on Pelee island. Kelley island and Ohio shore.

Cryptotaenia canadensis (L.) DC. (HONEWORT.)
Common at Point Pelee in damp rich woods and thickets, and on Pelee island. Ohio shore.

Zizia aurea (L.) Koch. (GOLDEN ALEXANDERS.)
Occasional at Point Pelee and on Pelee island. Kelley island.

Zizia cordata (Walt.) DC. (HEART-LEAVED ALEXANDERS.)
Frequent on Pelee island. Occasional along Detroit river.

Taenidia integerrima (L.) Drude. (YELLOW PIMPERNEL.)
Occasional at Point Pelee in dry open or slightly shaded ground and on Pelee island. Kelley and Put-in-Bay islands and Ohio shore.

Thaspium aureum Nutt. (MEADOW PARSNIP.)

Islands of Detroit river. (Maclagan.) Put-in-Bay island and Ohio shore.

Thaspium barbinode (Michx.) Nutt. (HAIRY-JOINTED MEADOW PARSNIP.)

Pelee island. Near Colchester and along Detroit river. Islands of Lake Erie and Ohio shore.

Thaspium barbinode (Michx.) Nutt., var. **angustifolium** Coult. and Rose. (HAIRY-JOINTED MEADOW PARSNIP.)

Pelee island. (Macoun.) Johnson and Mouse islands and Ohio shore.

Pastinaca sativa L. (PARSNIP.)

Near dwellings at Point Pelee as an escape from cultivation, and on Pelee island, Kelley island, and Ohio shore.

Heracleum lanatum Michx. (COW PARSNIP.)

Ohio shore.

Oxypolis rigidior (L.) Coult. and Rose. (COWBANE.)

Ohio shore.

Daucus Carota L. (CARROT.)

Occasional at Point Pelee about dwellings, in old fields, and on Pelee island. Ohio shore.

CORNACEAE (DOGWOOD FAMILY.)

Cornus canadensis L. (DWARF CORNEL.)

In damp shaded ground about Windsor. (F. P. Cravin.)

Cornus florida L. (FLOWERING DOGWOOD.)

Abundant on places along north shore of Lake Erie. Near Amherstburg. (Macoun.) Kelley island and Ohio shore.

Cornus circinata L'Her. (ROUND-LEAVED CORNEL.)

Occasional at Point Pelee on borders of woods and abundant on Pelee island. Hen and Kelley island and Ohio shore.

Cornus Amomum Mill. (SILKY CORNEL.)

Common at Point Pelee in damp open or shaded ground. Islands and Ohio shore.

Cornus asperifolia Michx. (ROUGH-LEAVED CORNEL.)
> Common at Point Pelee in dry open or slightly shaded ground, and along north shore of Lake Erie. Islands and Ohio shore.

Cornus Baileyi Coult. and Evans. (BAILEY'S CORNEL.)
> Occasional at Point Pelee and along north shore of Lake Erie.

Cornus stolonifera Michx. (RED-OSIER DOGWOOD.)
> Along north shore of Lake Erie. Ohio shore.

Cornus paniculata L'Her. (PANICLED CORNEL.)
> Common at Point Pelee on borders of woods and thickets, and on Pelee island.

Cornus alternifolia L. f. (ALTERNATE-LEAVED CORNEL.)
> Frequent at Point Pelee in open woods and on Pelee island. Ohio shore.

Nyssa sylvatica Marsh. (BLACK GUM.)
> Between Essex Centre and Leamington. (Macoun.) Near Colchester.

ERICACEAE (HEATH FAMILY.)

Chimaphila umbellata (L.) Nutt. (PRINCE'S PINE.)
> Ohio shore.

Pyrola elliptica Nutt. (SHIN LEAP.)
> Ohio shore.

Pyrola americana Sweet. (ROUND-LEAVED WINTERGREEN.)
> About Windsor. (F. P. Cravin.)

Monotropa uniflora L. (INDIAN PIPE.)
> Frequent about Windsor. (F. P. Cravin.) Ohio shore.

Gaultheria procumbens L. (WINTERGREEN.)
> Common in dry woods about Windsor.

Arctostaphylos Uva-ursi (L.) Spreng. (BEARBERRY.)
> On the west beach of Point Pelee acting as a sand binder. Pelee island.

Gaylussacia baccata (Wang.) C. Koch. (BLACK HUCKLEBERRY.)
> Common in dry woods about Windsor.

Vaccinium pennsylvanicum Lam. (EARLY SWEET BLUE-
BERRY.)

Common about Windsor.

Vaccinium vacillans Kalm. (LATE LOW BLUEBERRY.)

Common in woods about Windsor.

Vaccinium corymbosum L. (SWAMP BLUEBERRY.)

Common in swampy places about Windsor. (F. P.
Cravin.)

Vaccinium macrocarpon Ait. (LARGE CRANBERRY.)

Reported as formerly abundant in parts of the big marsh
at Point Pelee. Destroyed by drainage and fire.

PRIMULACEAE (PRIMROSE FAMILY.)

Lysimachia terrestris (L.) BSP. (BULB-BEARING LOOSE-
STRIFE.)

Occasional at Point Pelee in damp open ground. Bass
island and Ohio shore.

Lysimachia Nummularia L. (MONEYWORT.)

In damp places along north shore of Lake Erie. Middle
Bass island.

Lysimachia thyrsiflora L. (TUFTED LOOSESTRIFE.)

Frequent at Point Pelee in wet open spots. Ohio shore.

Steironema ciliatum (L.) Raf.

Common at Point Pelee in rich woods and thickets.
Islands and Ohio shore.

Steironema quadriflorum (Sims) Hitchc. (PRAIRIE MONEY-
WORT.)

Islands of Detroit river and near Sandwich. (Maclagan.)

OLEACEAE (OLIVE FAMILY.)

Fraxinus americana L. (WHITE ASH.)

Frequent at Point Pelee in rich ground with other trees.
Islands and Ohio shore.

Fraxinus pennsylvanica Marsh. (RED ASH.)

Occasional at Point Pelee in rich ground with other trees.
Islands and Ohio shore. Along shore of Lake St. Clair.

Fraxinus quadrangulata Michx. (BLUE ASH.)

    In dry ground at Point Pelee with other trees along "the narrows." Abundant on Pelee island. On the other Lake Erie islands and Ohio shore.

Fraxinus nigra Marsh. (BLACK ASH.)

    Occasional at Point Pelee and usually small in wet woods with other trees. Islands and Ohio shore.

Syringa vulgaris L. (COMMON LILAC.)

    Noticed as an escape on Pelee island. Well established on Kelley island.

Ligustrum vulgare L. (PRIVET.)

    Ohio shore.

GENTIANACEAE (GENTIAN FAMILY.)

Gentiana crinita Froel. (FRINGED GENTIAN.)

    Abundant in damp spots along Lake Erie shore. Ohio shore.

Gentiana quinquefolia L. (STIFF GENTIAN.)

    Along Detroit river. (Maclagan.)

Gentiana Andrewsii Griseb. (CLOSED GENTIAN.)

    In damp spots along Lake Erie shore and Detroit river. Frequent about Windsor. (F. P. Cravin.) Ohio shore.

APOCYNACEAE (DOGBANE FAMILY.)

Vinca minor L. (COMMON PERIWINKLE.)

    Kelley and Middle Bass islands and Ohio shore.

Apocynum androsaemifolium L. (SPREADING DOGBANE.)

    Frequent at Point Pelee in dry open ground and on Pelee island. Put-in-Bay and Middle Bass islands and Ohio shore.

Apocynum cannabinum L. (INDIAN HEMP.)

    Frequent at Point Pelee in damp gravelly ground. Islands and Ohio shore.

Apocynum cannabinum L., var. **pubescens** (R. Br.) DC. (VELVET DOGBANE.)

    In blown up sand at Point Pelee. (Macoun.)

MICROCOPY RESOLUTION TEST CHART

(ANSI and ISO TEST CHART No. 2)

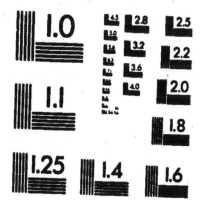

1653 East Main Street
Rochester, New York   14609   USA
(716) 482 - 0300 - Phone
(716) 288 - 5989 - Fax

ASCLEPIADACEAE (MILKWEED FAMILY.)

Asclepias tuberosa L. (BUTTERFLY-WEED.)
>Occasional at Point Pelee in dry open ground. Put-in-Bay and North Bass islands and Ohio shore.

Asclepias purpurascens L. (PURPLE MILKWEED.)
>Along Detroit river. (Maclagan.) Ohio shore.

Asclepias incarnata L. (SWAMP MILKWEED.)
>Common at Point Pelee in damp open places. Islands and Ohio shore.

Asclepias syriaca L. (COMMON MILKWEED.)
>Common at Point Pelee in dry open ground and old cultivated fields. Islands and Ohio shore.

Asclepias phytolaccoides Pursh. (POKE MILKWEED.)
>Common at Point Pelee. Occasional at Pelee island. Islands of Detroit river and on the mainland. (Maclagan.) Put-in-Bay island and Ohio shore.

Asclepias verticillata L. (WHORLED MILKWEED.)
>Ohio shore.

Acerates floridana (Lam.) Hitchc. (FLORIDA MILKWEED.)
>Along Detroit river. (Maclagan.) Near Sandwich. (Macoun.)

Acerates viridiflora Ell. (GREEN MILKWEED.)
>Frequent at Point Pelee on and near the upper sandy beach. Ohio shore.

Acerates viridiflora Ell., var. **lanceolata** (Ives) Gray. (GREEN MILKWEED.)
>Point Pelee. (Macoun.) Ohio shore.

CONVOLVULACEAE (CONVOLVULUS FAMILY.)

Ipomoea hederacea Jacq. (IVY-LEAVED MORNING GLORY.)
>Noticed at Point Pelee as an escape in sandy ground.

Ipomoea purpurea (L.) Roth. (COMMON MORNING GLORY.)
>In sandy ground near dwellings at Point Pelee. Put-in-Bay and North Bass islands and Ohio shore.

Ipomoea pandurata (L.) G.F.W. Mey. (WILD POTATO-VINE.)
>Very abundant at Point Pelee in spots in dry ground, and white with flowers, August 20, 1910. Ohio shore.

Convolvulus sepium L.   (HEDGE BINDWEED.)
>   Frequent at Point Pelee in damp open ground.   Islands
>   and Ohio shore.

Convolvulus sepium L., var. **pubescens** (Gray) Fernald.
(TRAILING BINDWEED.)
>   Ohio shore.

Convolvulus arvensis L.   (FIELD BINDWEED.)
>   Near dwellings and in old fields at Point Pelee.   Islands.

Cuscuta obtusiflora HBK.   (SMARTWEED DODDER.)
>   Ohio shore.

Cuscuta Cephalanthi Engelm.   (BUTTON-BUSH DODDER.)
>   Put-in-Bay island.

Cuscuta indecora Chois.   (PRETTY DODDER.)
>   Ohio shore.

Cuscuta Gronovii Willd.   (LOVE-VINE.)
>   Frequent at Point Pelee in open or shaded ground
>   especially on edge of marsh along "the narrows."   Com-
>   mon on the islands of Lake Erie.

POLEMONIACEAE (POLEMONIUM FAMILY.)

Phlox pilosa L.   (DOWNY PHLOX.)
>   Near Amherstburg in dry woods. (Macoun.)   Along
>   Detroit river. (Maclagan.)   Ohio shore.

Phlox divaricata L.   (BLUE PHLOX.)
>   Common at Point Pelee in rich open woods.   Islands and
>   Ohio shore.

Phlox subulata L.   (GROUND PINK.)
>   Ohio shore.

Polemonium reptans L.   (GREEK VALERIAN.)
>   Ohio shore.

HYDROPHYLLACEAE (WATERLEAF FAMILY.)

Hydrophyllum virginianum L.   (VIRGINIA WATERLEAF.)
>   Islands of Lake Erie except Kelley and Put-in-Bay is-
>   lands.   Ohio shore.

Hydrophyllum appendiculatum Michx. (APPENDAGED
WATERLEAF.)

Common at Point Pelee in damp shaded ground. Along
Detroit river. (Maclagan.) Islands of Lake Erie.

Phacelia Purshii Buckley. (PURSH'S PHACELIA.)

Chicken, Johnson, and Kelley islands and Ohio shore.

BORAGINACEAE (BORAGE FAMILY.)

Cynoglossum officinale L. (COMMON HOUND'S TONGUE.)

Common at Point Pelee in open ground near dwellings,
in cultivated fields and open woods. Islands and Ohio
shore.

Lappula virginiana (L.) Greene. (BEGGAR'S LICE.)

Occasional at Point Pelee in rich shaded ground. Kelley
and Put-in-Bay islands and Ohio shore.

Lappula echinata Gilibert. (EUROPEAN STICKSEED.)

Occasional at Point Pelee along the roads and on Pelee
island. Kelley and Middle Bass islands and Ohio
shore.

Myosotis virginica (L.) BSP. (EARLY SCORPION-GRASS.)

Put-in-Bay island.

Mertensia virginica (L.) Link. (BLUEBELLS.)

Johnson, Kelley, and North Bass islands and Ohio shore.

Lithospermum arvense L. (CORN GROMWELL.)

As a weed at Point Pelee about dwellings and in cultivated fields. Islands and Ohio shore.

Lithospermum Gmelini (Michx.) Hitchc. (HAIRY PUC-
COON.)

Frequent at Point Pelee in open sandy ground on and
near the sandy upper beach and about Windsor.
Islands and Ohio shore.

Lithospermum canescens (Michx.) Lehm. HOARY PUC-
COON.)

Along Detroit river. (Maclagan.) Ohio shore.

Lithospermum angustifolium Michx. (NARROW-LEAVED
PUCCOON.)

Occasional at Point Pelee in open sandy ground.

Onosmodium hispidissimum Mackenzie. (SHAGGY FALSE
GROMWELL.)
 Johnson islands and Ohio shore.
Echium vulgare L. (BLUE-WEED.)
 Common along railways. (F. P. Cravin.)

VERBENACEAE (VERVAIN FAMILY.)

Verbena urticaefolia L. (WHITE VERVAIN.)
 In dryish open ground at Point Pelee, but apparently
 infrequent. Islands and Ohio shore.
Verbena angustifolia Michx. (NARROW-LEAVED VERVAIN.)
 Noticed as frequent on Pelee island. Kelley island
 and Ohio shore.
Verbena hastata L. (BLUE VERVAIN.)
 Often very abundant at Point Pelee in damp open ground.
 Islands and Ohio shore.
Lippia lanceolata Michx. (FOG-FRUIT.)
 Wet places near Leamington. (Macoun.) Johnson and
 Put-in-Bay islands and Ohio shore.

LABIATAE (MINT FAMILY.)

Teucrium canadense L. (COMMON GERMANDER.)
 Frequent at Point Pelee in damp open ground. Islands
 and Ohio shore.
Teucrium occidentale Gray. (HAIRY GERMANDER.)
 Occasional at Point Pelee in the drier parts of the big
 marsh. Ohio shore.
Isanthus brachiatus (L.) BSP. (FALSE PENNYROYAL.)
 Kelley island and common on Ohio shore.
Scutellaria lateriflora L. (MAD-DOG SKULLCAP.)
 Common at Point Pelee in damp rich woods. Islands
 and Ohio shore.
Scutellar .ricolor Nutt. (HEART-LEAVED SKULLCAP.)
 Johnson and Put-in-Bay islands and Ohio shore.

Scutellaria galericulata L. (MARSH SKULLCAP.) (HOODED
WILLOW-HERB.)

> Common at Point Pelee in and about the big marsh and
> on Pelee island. Put-in-Bay and Middle Bass islands
> and Ohio shore.

Scutellaria parvula Michx. (SMALL SKULLCAP.)

> Abundant in spots on Pelee island in dry ground. Kelley
> island and Ohio shore.

Marrubium vulgare L. (COMMON HOREHOUND.)

> Frequent at Point Pelee in dry ground near dwellings.
> Islands and Ohio shore.

Agastache nepetoides (L.) Ktze. (CATNIP GIANT HYSSOP.)

> Frequent at Point Pelee in shaded ground. Along
> Detroit river. (Maclagan.) Kelley and Johnston is-
> lands and Ohio shore.

Nepeta Cataria L. (CATNIP.)

> Common at Point Pelee near dwellings and in old fields.
> Islands and Ohio shore.

Nepeta hederacea (L.) Trevisan. (GROUND IVY.)

> Common at Point Pelee and appearing like a native
> plant. Islands and Ohio shore.

Prunella vulgaris L. (HEAL-ALL.)

> Common at Point Pelee in damp open ground. Islands
> and Ohio shore.

Physostegia virginiana (L.) Benth. (FALSE DRAGON HEAD.)

> In damp sandy ground along Lake Erie shore. Put-in-
> Bay and Middle Bass islands and Ohio shore.

Lamium amplexicaule L. (HENBIT.)

> Occasional at Point Pelee as a weed in gardens. (Wal-
> lace Tilden.) Islands and Ohio shore.

Leonurus Cardiaca L. (COMMON MOTHERWORT.)

> Common at Point Pelee about dwellings and in old
> cultivated fields. Islands and Ohio shore.

Stachys tenuifolia Willd., var. **aspera** (Michx.) Fernald.
(ROUGH HEDGE NETTLE.)

> Middle Bass and North Bass islands and Ohio shore.

Stachys palustris L. (WOUNDWORT.)
> Frequent at Point Pelee in and about the big marsh, and on Pelee island.

Monarda didyma L. (BEE BALM.)
> Frequent about Windsor. (F. P. Cravin.)

Monarda fistulosa L. (WILD BERGAMOT.)
> At Point Pelee. (Macoun.) Islands and Ohio shore.

Monarda mollis L. (PALE WILD BERGAMOT.)
> Frequent at Point Pelee in dry open or slightly shaded ground. Islands and Ohio shore.

Blephilia ciliata (L.) Raf. (DOWNY BLEPHILIA.)
> Plentiful on Pelee island in open or shaded ground. Johnson, Kelley, and Put-in-Bay islands and Ohio shore.

Blephilia hirsuta (Pursh) Benth. (WOOD MINT.)
> Ohio shore.

Hedeoma pulegioides (L.) Pers. (AMERICAN PENNYROYAL.)
> Along Detroit river. (Maclagan.) Islands and Ohio shore.

Melissa officinalis L. (COMMON BALM.)
> Put-in-Bay island.

Satureja glabra (Nutt.) Fernald. (LOW CALAMINT.)
> Ohio shore.

Satureja hortensis L. (SUMMER SAVORY.)
> Ohio shore.

Satureja vulgaris (L.) Fritsch. (*Calamintha Clinopodium* Benth.) (FIELD BASIL.)
> Occasional at Point Pelee in dry open or slightly shaded ground. Islands and Ohio shore.

Pycnanthemum virginianum (L.) Durand and Jackson. (VIRGINIAN MOUNTAIN MINT.)
> Common at Point Pelee in the drier part of the big marsh. Put-in-Bay island and Ohio shore.

Pycnanthemum pilosum Nutt. (HAIRY MOUNTAIN MINT.)
> In dry shaded ground at Point Pelee, but apparently infrequent. Ohio shore.

Lycopus virginicus L. (PURPLE BUGLE WEED.)
> Frequent at Point Pelee in moist open places. Islands and Ohio shore.

Lycopus uniflorus Michx. (COMMON BUGLE WEED.)
Frequent at Point Pelee in low moist ground.

Lycopus rubellus Moench. (STALKED WATER HOREHOUND.)
In damp grassy places at Point Pelee. Apparently
infrequent. Islands and Ohio shore.

Lycopus americauus Muhl. (CUT-LEAVED WATER HORE-
HOUND.)
Common and often abundant at Point Pelee in damp
open places, especially in and about the big marsh.
Islands and Ohio shore.

Mentha spicata L. (SPEARMINT.)
Along roads on Pelee island. Put-in-Bay island and
Ohio shore.

Mentha piperita L. (PEPPERMINT.)
Occasional at Point Pelee in damp places near dwellings
and in damp sand on the beach.

Mentha arvensis L., var. **canadensis** (L.) Briquet. (AMERI-
CAN WILD MINT.)
Frequent at Point Pelee in damp open or slightly shaded
ground. Common on the islands and Ohio shore.

Collinsonia canadensis L. (HORSE BALM.)
Along Detroit river. (Maclagan.)

SOLANACEAE (NIGHTSHADE FAMILY.)

Solanum Dulcamara L. (NIGHTSHADE.)
Common at Point Pelee in damp open or shaded ground.
Islands and Ohio shore.

Solanum nigrum L. (COMMON NIGHTSHADE.)
Frequent at Point Pelee in rich open or shaded ground,
and a weed in gardens. Islands and Ohio shore.

Solanum carolinense L. (HORSE NETTLE.)
Ohio shore.

Solanum rostratum Dunal. (BUFFALO BUR.)
Put-in-Bay island and Ohio shore.

Physalis pruinosa L. (STRAWBERRY TOMATO.)
Kelley island.

Physalis heterophylla Nees. (CLAMMY GROUND CHERRY.)
    Common at Point Pelee in open sandy ground. Islands and Ohio shore.

Physalis heterophylla Nees, var. **ambigua** (Gray) Rydb. (CLAMMY GROUND CHERRY.)
    Ohio shore.

Physalis heterophylla Nees, var. **nyctaginea** (Dunal) Rydb. (CLAMMY GROUND CHERRY.)
    Ohio shore.

Physalis subglabrata Mackenzie and Bush. (GLABRATE GROUND CHERRY.)
    Along Detroit river.

Physalis lanceolata Michx. (PRAIRIE GROUND CHERRY.)
    Kelley island and Ohio shore.

Lycopersicon esculentum Mill. (TOMATO.)
    Said to have escaped and become well established on Kelley and Put-in-Bay islands.

Lycium halimifolium Mill. (COMMON MATRIMONY VINE.)
    Occasional at Point Pelee as an escape and on Pelee island. Kelley island.

Hyoscyamus niger L. (BLACK HENBANE.)
    Along Detroit river. (Maclagan.)

Datura Stramonium L. (STRAMONIUM.)
    Ohio shore.

Datura Tatula L. (PURPLE STRAMONIUM.)
    Occasional at Point Pelee as a weed about dwellings and in waste places.

LARIACEAE (FIGWORT FAMILY.)

Verbascum ▃▃▃▃▃us L. (COMMON MULLEIN.)
    Frequent at Point Pelee as a weed in cultivated grounds.

Verbascum Blattaria L. (Moth Mullein.)
    Frequent along roads on Pelee island. Along Detroit river and near Windsor.

Verbascum Lychnitis L. (WHITE MULLEIN.)
    Roadside near Sandwich. (Macoun.)

Linaria vulgaris Hill. (BUTTER AND EGGS.)

Frequent at Point Pelee about dwellings as a weed.
Islands and Ohio shore.

Scrophularia leporella Bicknell. (HARE FIGWORT.)

Frequent at Point Pelee on borders of the big marsh.
Islands and Ohio shore.

Pentstemon hirsutus (L.) Willd. (HAIRY BEARD-TONGUE.)

Frequent at Point Pelee in dry open or slightly shaded
ground. Islands and Ohio shore.

Chelone glabra L. (TURTLEHEAD.)

Frequent in damp ground along Lake Erie shore.

Mimulus ringens L. (SQUARE-STEMMED MONKEY FLOWER.)

In ditches and damp places about the big marsh at Point
Pelee and on Pelee island. Bass island and Ohio shore.

Conobea multifida (Michx.) Benth. (CONOBEA.)

On the extreme southern point of Pelee island. (Macoun.)
Kelley island and Ohio shore.

Ilysanthes dubia (L.) Barnhart. (LONG-STALKED FALSE
PIMPERNEL.)

In damp sandy ground along Lake Erie shore. Ohio
shore.

Veronica virginica L. (CULVER'S-ROOT.)

Islands of Detroit river. (Maclagan.) Ohio shore.

Veronica Anagallis-aquatica L. (WATER SPEEDWELL.)

Kelley island.

Veronica officinalis L. (COMMON SPEEDWELL.)

Ohio shore.

Veronica serpyllifolia L. (THYME-LEAVED SPEEDWELL.)

Frequent at Point Pelee in open grassy places and on
Pelee island. Put-in-Bay island.

Veronica peregrina L. (NECKWEED.)

Occasional at Point Pelee as a weed in gardens and old
fields and on Pelee island. Put-in-Bay, North Bass,
and Rattlesnake islands and Ohio shore.

Veronica arvensis L. (CORN SPEEDWELL.)

In open grassy places at Point Pelee. Common on the
islands and Ohio shore.

Seymeria macrophylla Nutt. (MULLEIN FOXGLOVE.)
Ohio shore.

Gerardia pedicularia L. (FERN-LEAVED FOXGLOVE.)
In dry open ground about Windsor. (F. P. Cravin.)

Gerardia virginica (L.) BSP. (SMOOTH FALSE FOXGLOVE.)
Along Detroit river. (Maclagan.) Ohio shore.

Gerardia auriculata Michx. (AURICLED GERARDIA.)
Ohio shore.

Gerardia purpurea L. (PURPLE GERARDIA.)
About Windsor. (F. P. Cravin.) Ohio shore.

Gerardia paupercula (Gray) Britton. (SMALL-FLOWERED GERARDIA.)
Frequent at Point Pelee about the big marsh, and on Pelee island.

Gerardia tenuifolia Vahl. (SLENDER GERARDIA.)
In damp places along Lake Erie shore. Along Detroit river. (Maclagan) Kelley island and Ohio shore.

Castilleja coccinea (L.) Spreng. (SCARLET PAINTED CUP.)
Common about Windsor. (F. P. Cravin.) Ohio shore.

Pedicularis canadensis L. (COMMON LOUSEWORT.)
Frequent at Point Pelee in shaded ground, and on Pelee island. Kelley and Put-in-Bay islands and Ohio shore.

Pedicularis lanceolata Michx. (SWAMP LOUSEWORT.)
In damp open places along Lake Erie shore. Along Detroit river. (Maclagan.)

LENTIBULARIACEAE (BLADDERWORT FAMILY.)

Utricularia vulgaris L. ar **americana** Gray. (GREATER BLADDERWORT.)
Common in shallow water about Windsor. (F. P. Cravin.) Ohio shore.

Utricularia gibba L. (HUMPED BLADDERWORT.)
Ohio shore.

OROBANCHACEAE (Broom-rape Family.)

Conopholis americana (L. f.) Wallr. (Squaw-root.)
Put-in-Bay island.
Orobanche uniflora L. (One-flowered Cancer-root.)
Ohio shore.

BIGNONIACEAE (Bignonia Family.)

Tecoma radicans (L.) Juss. (Trumpet Creeper.)
Common on Pelee island. The other Lake Erie islands, and Ohio shore.
Catalpa speciosa Warder. (Catawba Tree.)
Often planted and apparently spreading near Kingsville.
Catalpa bignonioides Walt. (Catalpa.)
Planted and escaping along Lake Erie shore.

ACANTHACEAE (Acanthus Family.)

Dianthera americana L. (Dense-flowered Water Willow.)
Put-in-Bay and Middle Bass islands and Ohio shore.

PHRYMACEAE (Lopseed Family.)

Phryma Leptostachya L. (Lopseed.)
Frequent at Point Pelee in rich woods and thickets. Kelley and Put-in-Bay islands and Ohio shore.

PLANTAGINACEAE (Plantain Family.)

Plantago cordata Lam. (Hart-leaved Plantain.)
Along Detroit river. (Maclagan.) Near Amherstburg. (Macoun.)
Plantago major L. (Common Plantain.)
Frequent at Point Pelee about dwellings and in waste places. Islands and Ohio shore.

Plantago Rugelii Dene. (RUGEL'S PLANTAIN.)

In fields and pastures at Point Pelee. Islands and Ohio shore.

Plantago lanceolata L. (ENGLISH PLANTAIN.)

Occasional at Point Pelee as a weed along roads and in cultivated grounds and on Pelee island.

Plantago aristata Michx. (LARGE-BRACTED PLANTAIN.)

Roadsides near Windsor. (Macoun.)

RUBIACEAE (MADDER FAMILY.)

Galium Aparine L. (CLEAVERS.)

Common at Point Pelee in rich shaded ground. Islands and Ohio shore.

Galium pilosum Ait. (HAIRY BEDSTRAW.)

Point Pelee. (Macoun.) Along Detroit river and Ohio shore.

Galium circaezans. Michx. (WILD LIQUORICE.)

Occasional at Point Pelee in dry shaded ground, and on Pelee island. Put-in-Bay, Middle Bass, and Rattlesnake islands, and Ohio shore.

Galium lanceolatum Torr. (WILD LIQUORICE.)

In dry woods about Windsor. (F. P. Cravin.)

Galium boreale L. (NORTHERN BEDSTRAW.)

Frequent on Pelee island. Kelley island and Ohio shore.

Galium trifidum L. (SMALL BEDSTRAW.)

Common at Point Pelee on the borders of the big marsh, especially along "the narrows," and on Pelee island. Put-in-Bay and Middle Bass islands and Ohio shore.

Galium tinctorium L. (STIFF MARSH BEDSTRAW.)

Frequent in damp open places at Point Pelee, and on Pelee island. Ohio shore.

Galium concinnum T. and G. (SHINING BEDSTRAW.)

Ohio shore.

Galium asprellum Michx. (ROUGH BEDSTRAW.)

Occasional at Point Pelee in wet bushy places. Islands and Ohio shore.

Galium triflorum Michx. (SWEET-SCENTED BEDSTRAW.)

Common at Point Pelee in rich shaded ground, and on Pelee island. Rattlesnake island.

Cephalanthus occidentalis L. (BUTTONBUSH.)

Abundant at Point Pelee on borders of the big marsh, especially along "the narrows." Islands and Ohio shore.

Houstonia longifolia Gaertn. (SLENDER-LEAVED HOUSTONIA.)

Rattlesnake and Put-in-Bay islands and Ohio shore.

Houstonia ciliolata Torr. (FRINGED HOUSTONIA.)

Common on the Ohio shore.

CAPRIFOLIACEAE (HONEYSUCKLE FAMILY.)

Diervilla Lonicera Mill. (BUSH HONEYSUCKLE.)

In dry ground about Windsor. (F. P. Cravin.)

Lonicera glaucescens Rydb. (DOUGLAS HONEYSUCKLE.)

Frequent at Point Pelee in dry open or shaded ground. Islands and Ohio shore.

Symphoricarpos racemosus Michx. (SNOWBERRY.)

Abundant at Point Pelee in dry shaded ground. Islands and Ohio shore.

Symphoricarpos racemosus Michx., var. **pauciflorus** Robbins. (LOW SNOWBERRY.)

Islands and Ohio shore.

Triosteum perfoliatum L. (TINKER'S WEED.)

About Windsor. (F. P. Cravin.)

Viburnum acerifolium L. (DOCKMACKIE.)

Occasional at Point Pelee in dry open or shaded ground. Put-in-Bay island.

Viburnum pubescens (Ait.) Pursh. (DOWNY ARROW-WOOD.)

Occasional at Point Pelee in thickets, and on Pelee island. Kelley and Put-in-Bay islands and Ohio shore.

Viburnum dentatum L. (ARROW-WOOD.)

"In thickets" at Point Pelee. (Macoun.) Not noticed in 1910–11.

Viburnum Lentago L. (NANNYBERRY.)

    Frequent at Point Pelee in damp open woods, and on Pelee island. Kelley and Middle Bass islands and Ohio shore.

Viburnum prunifolium L. (BLACK HAW.)

    Common on Ohio shore.

Sambucus canadensis L. (COMMON ELDER.)

    Frequent at Point Pelee in rich open or slightly shaded ground. Islands and Ohio shore.

VALERIANACEAE (VALERIAN FAMILY.)

Valeriana pauciflora Michx. (LARGE-FLOWERED VALERIAN.)

    Ohio shore.

DIPSACACEAE (TEASEL FAMILY.)

Dipsacus sylvestris Huds. (WILD TEASEL.)

    In open ground and on roadsides along Lake Erie shore. Kelley island and Ohio shore.

CUCURBITACEAE (GOURD FAMILY.)

Sicyos angulatus L. (ONE-SEEDED BUR CUCUMBER.)

    Occasional at Point Pelee as a native plant in damp thickets, and as a weed in yards and fields. Green, Rattlesnake, and Put-in-Bay islands and Ohio shore.

Echinocystis lobata (Michx.) T. and G. (WILD BALSAM APPLE.)

    Occasional at Point Pelee as an escape near dwellings. Islands and Ohio shore.

CAMPANULACEAE (BLUEBELL FAMILY.)

Specularia perfoliata (L.) A. DC. (VENUS'S LOOKING-GLASS.)

    Pelee island. Kelley and Put-in-Bay islands and Ohio shore.

Campanula americana L. (TALL BELLFLOWER.)

Frequent at Point Pelee in damp shaded ground, especially on the borders of the big marsh along "the narrows." Common on the islands and Ohio shore.

Campanula rotundifolia L. (HAREBELL.)

Occasional on bluffy shores of Lake Erie and on rocky shores of islands, except Kelley island. Ohio shore.

Campanula aparinoides Pursh. (MARSH BELLFLOWER.)

Common at Point Pelee in and about the big marsh. Islands and Ohio shore.

LOBELIACEAE (LOBELIA FAMILY.)

Lobelia cardinalis L. (CARDINAL FLOWER.)

In rich open woods at Point Pelee. Islands and Ohio shore.

Lobelia siphilitica L. (GREAT LOBELIA.)

Frequent at Point Pelee in and about the big marsh. Kelley, Middle Bass, and North Bass islands, and Ohio shore.

Lobelia spicata Lam. (PALE SPIKED LOBELIA.)

Frequent on Pelee island.

Lobelia Kalmii L. (BROOK LOBELIA.)

Islands and Ohio shore.

Lobelia inflata L. (INDIAN TOBACCO.)

Occasional along Lake Erie shore. Put-in-Bay island.

COMPOSITAE (COMPOSITE FAMILY.)

Vernonia noveboracensis Willd. (NEW YORK IRONWEED.)

Pelee island. (Macoun.) (See Gray's New Manual of Botany, Illustrated, p. 780.)

Vernonia fasciculata Michx. (WESTERN IRONWEED.)

Ohio shore.

Vernonia altissima Nutt. (TALL IRONWEED.)

Near Essex Centre. Contributions from the Herbarium of the Geological Survey of Canada, IV, p. 202. Ohio shore.

Vernonia illinoensis Gleason. (ILLINOIS IRONWEED.)

Frequent at Point Pelee in damp open ground. On flat open ground about Lake St. Clair. Kelley island.

Eupatorium purpureum L. (JOE-PYE WEED.)

Ohio shore.

Eupatorium purpureum L., var. **maculatum** (L.) Darl.

Ohio shore.

Eupatorium altissimum L. (TALL THOROUGHWORT.)

Johnson island and Ohio shore.

Eupatorium sessilifolium L. (UPLAND BONESET.)

Ohio shore.

Eupatorium perfoliatum L. (BONESET.)

Common at Point Pelee in damp open ground. Islands and Ohio shore.

Eupatorium urticaefolium Reichard (*E. ageratoides* L.f.) (WHITE SNAKEROOT.)

Rattlesnake island and Ohio shore.

Liatris scariosa Willd. (LARGE BUTTON SNAKEROOT.)

Near Leamington and along Detroit river. (Burgess.) Ohio shore.

Liatris spicata (L.) Willd. (DENSE BUTTON SNAKEROOT.)

Ohio shore.

Solidago caesia L. (BLUE-STEMMED GOLDENROD.)

Abundant in spots along the Lake Erie shore. Islands and Ohio shore.

Solidago latifolia L. (ZIGZAG GOLDENROD.)

Kelley, Green, and Rattlesnake islands, and Ohio shore.

Solidago hispida Muhl. (HAIRY GOLDENROD.)

Frequent at Point Pelee in dry open or slightly shaded ground. Islands and Ohio shore.

Solidago speciosa Nutt. (SHOWY GOLDENROD.)

Ohio shore.

Solidago patula Muhl. (ROUGH-LEAVED GOLDENROD.)

Kelley island.

Solidago juncea Ait. (EARLY GOLDENROD.)

Occasional at Point Pelee in dry open ground, and on Pelee island. Ohio shore.

Solidago juncea Ait., var. **scabrella** (T. and G.) Gray. (EARLY GOLDENROD.)

Near Leamington. (Macoun.)

Solidago ulmifolia Muhl. (ELM-LEAVED GOLDENROD.)

Islands and Ohio shore.

Solidago rugosa Mill. (WRINKLE-LEAVED GOLDENROD.)

In dry open ground near Windsor.

Solidago nemoralis Ait. (FIELD GOLDENROD.)

Frequent at Point Pelee in dry open ground. Islands and Ohio shore.

Solidago canadensis L. (CANADA GOLDENROD.)

Common at Point Pelee in damp open ground. Islands and Ohio shore.

Solidago altissima L. (TALL GOLDENROD.)

Common at Point Pelee in damp open ground on borders of woods and thickets.

Solidago serotina Ait. (LATE GOLDENROD.)

Frequent at Point Pelee in thickets and borders of rich woods. Along Detroit river and Ohio shore.

Solidago rigida L. (STIFF GOLDENROD.)

On the islands of Detroit river. Middle Bass island and Ohio shore.

Solidago Riddellii Frank. (RIDDELL'S GOLDENROD.)

Ohio shore.

Solidago graminifolia (L.) Salisb. (BUSHY GOLDENROD.)

Common at Point Pelee and often abundant in damp open ground. Islands and Ohio shore.

Boltonia asteroides (L.) L'Her. (ASTER-LIKE BOLTONIA.)

Johnson and Put-in-Bay islands and Ohio shore.

Aster divaricatus L. (WHITE WOOD ASTER.)

Frequent about Windsor. (F. P. Cravin.)

Aster macrophyllus L. (LARGE-LEAVED ASTER.)

Occasional at Point Pelee in dry shaded ground. Put-in-Bay island.

Aster oblongifolius Nutt. (AROMATIC ASTER.)

About Windsor. (F. P. Cravin.)

Aster novae-angliae L. (NEW ENGLAND ASTER.)

    Occasional at Point Pelee in dryish open ground and on Pelee island. Common about Windsor. (F. P. Cravin.) Kelley and Put-in-Bay islands.

Aster azureus Lindl. (SKY-BLUE ASTER.)

    Occasional at Point Pelee in dry open or slightly shaded ground. Common about Windsor. (F. P. Cravin.) Ohio shore.

Aster Shortii Lindl. (SHORT'S ASTER.)

    Occasional at Point Pelee in dry ground. Islands and Ohio shore.

Aster undulatus L. (WAVY-LEAF ASTER.)

    Near Sandwich. (Maclagan.) Near Windsor. (Macoun.)

Aster Lowrieanus Porter, var., **lanceolatus** Porter. (LOWRIE'S ASTER.)

    In open woods along Lake Erie shore.

Aster sagittifolius Wedemeyer. (ARROW-LEAVED ASTER.)

    Common along Lake Erie shore in open or slightly shaded ground. About Windsor. (F. P. Cravin.) Common on the islands and Ohio shore.

Aster laevis L. (SMOOTH ASTER.)

    Occasional at Point Pelee in dry open or slightly shaded ground. Common about Windsor. (F. P. Cravin.) Ohio shore.

Aster polyphyllus Willd. (FAXON'S ASTER.)

    Put-in-Bay island and Ohio shore.

Aster ericoides L. (WHITE HEATH ASTER.)

    Occasional at Point Pelee in dry open ground. In sandy thickets near Windsor. (J. M. Macoun.) Hen island and Ohio shore.

Aster amethystinus Nutt. (AMETHYST ASTER.)

    About Windsor. (F. P. Cravin.)

Aster multiflorus Ait. (DENSE-FLOWERED ASTER.)

    In dry open ground along Lake Erie shore. Common near Windsor. Put-in-Bay island.

Aster dumosus L. (BUSHY ASTER.)

    About Windsor. (F. P. Cravin.)

Aster vimineus Lam. (SMALL WHITE ASTER.)
>About Windsor. (F. P. Cravin.)

Aster lateriflorus (L.) Britton. (CALICO ASTER.)
>Common along Lake Erie shore and about Windsor.

Aster Tradescanti L. (TRADESCANT'S ASTER.)
>Common at Point Pelee in and about the big marsh and
>in other wet places. Pelee island and Kelley island.

Aster paniculatus Lam. (PANICLED ASTER.)
>Common on the islands and Ohio shore.

Aster salicifolius Ait. (WILLOW ASTER.)
>Ohio shore.

Aster junceus Ait. (RUSH ASTER.)
>Abundant at Point Pelee in the big marsh.

Aster puniceus L. (RED-STALK ASTER.)
>Frequent at Point Pelee in swampy open places and along
>the Lake Erie shore.

Erigeron pulchellus Michx. (ROBIN'S PLANTAIN.)
>Common about Windsor.

Erigeron philadelphicus L. (PHILADELPHIA FLEABANE.)
>Occasional at Point Pelee in open places and fields.
>Islands.

Erigeron annuus (L.) Pers. (SWEET SCABIOUS.)
>Occasional at Point Pelee in rich open woods. Islands
>and Ohio shore.

Erigeron ramosus (Walt.) BSP. (DAISY FLEABANE.)
>Occasional at Point Pelee in dry open ground. Islands
>and Ohio shore.

Erigeron canadensis L. (HORSE-WEED.)
>A weed at Point Pelee near dwellings, in gardens and
>fields. Islands and Ohio shore.

Antennaria plantaginifolia (L.) Richards. (PLANTAIN-
LEAVED EVERLASTING.)
>Common on Kelley and Put-in-Bay islands and Ohio
>shore.

Antennaria neglecta Greene. (FIELD CAT'S-FOOT.)
>Ohio shore.

Gnaphalium polycephalum Michx. (COMMON EVERLAST-
ING.)

> Frequent at Point Pelee in dry open ground. Islands
> and Ohio shore.

Gnaphalium decurrens Ives. (EVERLASTING.)

> Ohio shore.

Gnaphalium uliginosum L. (LOW CUDWEED.)

> Ohio shore.

Inula Helenium L. (ELECAMPANE.)

> Along roads and in fields in Essex county. Ohio shore.

Polymnia canadensis L. (SMALL-FLOWERED LEAFCUP.)

> Abundant along south shore of Pelee island. Islands
> and Ohio shore.

Silphium terebinthinaceum Jacq. (PRAIRIE DOCK.)

> Near Amherstburg, and generally along Detroit river.
> Ohio shore.

Silphium trifoliatum L. (WHORLED ROSIN-WEED.)

> Near Amherstburg.

Silphium perfoliatum L. (CUP PLANT.)

> Islands in Detroit river. (Maclagan.) Margins of fields
> near Windsor. (Macoun.)

Ambrosia trifida L. (GREAT RAGWEED.)

> Occasional at Point Pelee as a weed in cultivated fields.
> Abundant on Pelee island. Common on the other
> islands and Ohio shore.

Ambrosia trifida L., var. **integrifolia** (Muhl.) T. and G.

> Usually growing with the preceding.

Ambrosia artemisiifolia L. (COMMON RAGWEED.)

> Common on Point Pelee as a weed in gardens and fields.
> Islands and Ohio shore.

Xanthium canadense Mill. (AMERICAN COCKLEBUR.)

> Occasional at Point Pelee in damp open ground. Islands
> and Ohio shore.

Xanthium commune Britton. (COMMON CLOTBUR.)

> Ohio shore.

Xanthium echinatum Murr. (BEACH COCKLEBUR.)

> Occasional at Point Pelee in sand on or near the beach,
> and on Pelee island.

Heliopsis helianthoides (L.) Sweet. (Ox-eye.)
>Occasional at Point Pelee in damp open ground. Islands and Ohio shore.

Eclipta alba (L.) Hassk. (Eclipta.)
>Ohio shore.

Rudbeckia triloba L. (Thin-leaved Cone-flower.)
>Ohio shore.

Rudbeckia hirta L. (Yellow Daisy.)
>Frequent in open ground along Lake Erie shore. Islands.

Rudbeckia laciniata L. (Tall Cone-flower.)
>Along Detroit river. Ohio shore.

Lepachys pinnata (Vent.) T. and G. (Gray-headed Cone-flower.)
>Near Amherstburg. Ohio shore.

Helianthus annuus L. (Common Sunflower.)
>Ohio shore.

Helianthus giganteus L. (Tall Sunflower.)
>Infrequent at Point Pelee, but common along Lake Erie shore. Islands and Ohio shore.

Helianthus divaricatus L. (Woodland Sunflower.)
>In dry shaded ground along Lake Erie shore. Islands and Ohio shore.

Helianthus hirsutus Raf. (Stiff-haired Sunflower.)
>Ohio shore.

Helianthus strumosus L. (Pale-leaved Wood Sunflower.)
>Occasional at Point Pelee in open woods and thickets. Ohio shore.

Helianthus strumosus L., var. **mollis** T. and G. (Pale-leaved Wood Sunflower.)
>Ohio shore.

Helianthus trachellifolius Mill. (Throatwort Sunflower.)
>Ohio shore.

Helianthus decapetalus L. (Thin-leaved Sunflower.)
>Ohio shore.

Helianthus tuberosus L. (Jerusalem Artichoke.)
>An occasional escape at Point Pelee. Apparently native along Lake Erie shore. Kelley and Put-in-Bay islands and Ohio shore.

Actinomeris alternifolia (L.) DC. (ACTINOMERIS.)
>Ohio shore.

Coreopsis tripteris L. (TALL COREOPSIS.)
>Islands in Detroit river. Islands and Ohio shore.

Bidens discoidea (T. and G.) Britton.
>Along Detroit river. (Maclagan.) Ohio shore.

Bidens frondosa L. (BEGGAR-TICKS.)
>Common at Point Pelee in damp open ground. Islands
>and Ohio shore.

Bidens comosa (Gray.) Wiegand. (LEAFY-BRACTED TICK-
SEED.)
>Frequent at Point Pelee in wet places in and about the
>big marsh. Ohio shore.

Bidens connata Muhl. (SWAMP BEGGAR-TICKS.)
>Common at Point Pelee in ditches and damp places about
>the big marsh. Ohio shore.

Bidens cernua L. (STICK-TIGHT.)
>Common at Point Pelee in and about the big marsh, and
>on Pelee island.

Bidens laevis (L.) BSP. (LARGER BUR-MARIGOLD.)
>Common at Point Pelee in and about the big marsh.
>Along Lake Erie shore. Islands and Ohio shore.

Bidens bipinnata L. (SPANISH NEEDLES.)
>North Bass island and Ohio shore.

Bidens trichosperma (Michx.) Britton, var. **tenuiloba** (Gray.)
Britton. (TALL TICKSEED SUNFLOWER.)
>Frequent at Point Pelee and in spots abundant in and
>about the big marsh. Islands in Detroit river. Pelee
>island and Kelley island.

Bidens aristosa (Michx.) Britton. (WESTERN TICKSEED SUN-
FLOWER.)
>Ohio shore.

Bidens Beckii Torr. (WATER MARIGOLD.)
>In very wet places along Detroit river. (Maclagan.)
>Ohio shore.

Actinea herbacea (Greene) Robinson. (STEMLESS PICRA-
DENIA.)
>Ohio shore.

Helenium autumnale L. (SWAMP SUNFLOWER.)

Occasional in damp places along Lake Erie shore. Ohio shore.

Achillea Millefolium L. (COMMON YARROW.)

Common at Point Pelee in open ground. Islands and Ohio shore.

Anth .mis Cotula L. (MAY-WEED.)

Frequent at Point Pelee about dwellings as a weed. Islands and Ohio shore.

Chrysanthemum Leucanthemum L., var. **pinnatifidum** Lecoq and Lamotte. (OX-EYE DAISY.)

Infrequent at Point Pelee. Put-in-Bay island and Ohio shore.

Chrysanthemum Parthenium (L.) Bernh. (FEVERFEW.)

Well established on Put-in-Bay island.

Chrysanthemum Balsamita L., var. **tanacetoides** Boiss. (COSTMARY.)

Pelee island and Ohio shore.

Tanacetum vulgare L. (COMMON TANSY.)

Established near dwellings at Point Pelee and on Pelee island. Kelley island.

Tanacetum vulgare L., var. **crispum** DC. (TANSY.)

Islands in Lake Erie.

Artemisia caudata Michx. (TALL WORMWOOD.)

Frequent at Point Pelee on the sandy beach. Islands and Ohio shore.

Artemisia biennis Willd. (BIENNIAL WORMWOOD.)

Occasional at Point Pelee as a weed in damp ground. Johnson, Middle Bass, and North Bass islands, and Ohio shore.

Calendula officinalis L. (POT MARIGOLD.)

Becoming naturalized on Put-in-Bay island.

Erechtites hieracifolia (L.) Raf. (FIREWEED.)

Frequent at Point Pelee bordering the big marsh. Islands and Ohio shore.

Senecio aureus L. (GOLDEN RAGWORT.)

Frequent about Windsor. (F. P. Cravin.)

Senecio Balsamitae Muhl. (BALSAM GROUNDSEL.)

Occasional at Point Pelee in dry open or slightly shaded ground and on Pelee island. Put-in-Bay island.

Senecio obovatus Muhl. (ROUND-LEAF SQUAW-WEED.)

Kelley island.

Arctium minus Bernh. (COMMON BUR   L.)

Common at Point Pelee as a weed. Islands and Ohio shore.

Cirsium lanceolatum (L.) Hill. (COMMON THISTLE.)

Occasional at Point Pelee about dwellings and in old fields. Islands and Ohio shore.

Cirsium discolor (Muhl.) Spreng. (FIELD THISTLE.)

Ohio shore.

Cirsium altissimum (L.) Spreng. (TALL THISTLE.)

Kelley island and Ohio shore.

Cirsium muticum Michx. (SWAMP THISTLE.)

Occasional at Point Pelee in wet open or slightly shaded places, and along Lake Erie shore.

Cirsium arvense (L.) Scop. (CANADA THISTLE.)

Frequent at Point Pelee in fields and pastures. Islands and Ohio shore.

Centaurea Cyanus L. (BLUEBOTTLE.)

Kelley island.

Cichorium Intybus L. (CHICORY.)

Occasional at Point Pelee in old fields, and on Pelee island. Kelley and Middle Bass islands and Ohio shore.

Krigia amplexicaulis Nutt. (CYNTHIA.)

Occasional on Pelee island. On islands of Detroit river. (Maclagan.) Kelley island and Ohio shore.

Tragopogon porrifolius L. (SALSIFY.)

Pelee island and on the other islands.

Tragopogon pratensis L. (GOAT'S BEARD.)

In towns and along railways.

Taraxacum officinale Weber. (COMMON DANDELION.)

Abundant at Point Pelee as a weed in yards, pastures, and cultivated fields. Islands and Ohio shore.

Taraxacum erythrospermum Andrz. (RED-SEEDED DANDE-
LION.)

    Ohio shore.

Sonchus asper L. Hill. (SPINY-LEAVED SOW THISTLE.)

    A weed at Point Pelee in gardens and waste places.
Islands and Ohio shore.

Lactuca scariola L. (PRICKLY LETTUCE.)

    Frequent at Point Pelee on banks of ditches. Islands
and Ohio shore.

Lactuca scariola L., var. **integrata** Gren. and Godr. (PRICKLY
LETTUCE.)

    Common at Point Pelee on banks of ditches and in waste
places. Islands and Ohio shore.

Lactuca canadensis L. (WILD LETTUCE.)

    Occasional at Point Pelee in rich open or shaded ground.
Islands and Ohio shore.

Lactuca canadensis L., var. **montana** Britton. (WILD
LETTUCE.)

    Occasional at Point Pelee in open woods.

Lactuca hirsuta Muhl. (HAIRY WOOD-LETTUCE.)

    In thickets near Leamington. (Macoun.)

Lactuca villosa Jacq. (HAIRY-VEINED BLUE LETTUCE.)

    Ohio shore.

Lactuca floridana (L.) Gaertn. (FLORIDA LETTUCE.)

    Islands of Detroit river. (Maclagan.) Put-in-Bay and
Green islands.

Lactuca spicata (Lam.) Hitchc. (TALL BLUE LETTUCE.)

    Frequent at Point Pelee on shaded borders of the big
marsh, especially along "the narrows," and on Pelee
island. Kelley and Put-in-Bay islands and Ohio shore.

Prenanthes racemosa Michx. (GLAUCOUS WHITE LETTUCE.)

    Along Detroit river. (Maclagan.)

Prenanthes racemosa Michx., var. **pinnatifida** Gray. (GLAU-
COUS WHITE LETTUCE.)

    Near Windsor. (Wm. Scott.)

Prenanthes alba L. (WHITE LETTUCE.)

    Occasional at Point Pelee in rich open or shaded ground,
and along Lake Erie shore. Islands and Ohio shore.

Prenanthes altissima L.   (TALL WHITE LETTUCE.)
   Put-in-Bay islands and Ohio shore.
Hieracium venosum L.   (RATTLESNAKE-WEED.)
   Along Detroit river.   (Maclagan.)
Hieracium scabrum Michx.   (ROUGH HAWKWEED.)
   Ohio shore.
Hieracium Gronovii L.   (GRONOVIUS' HAWKWEED.)
   Near Sandwich.   (Maclagan.)
Hieracium canadense Michx.   (CANADA HAWKWEED.)
   Ohio shore.

INDEX.

A.—Con.

B.—*Con.*

C.

113

C.—Con.

C.—Con.

D.

E.

E.—Con.

118

L.

L.—Con.

M.

N.

O.

P.

P.—*Con.*

P.—*Con.*

P.—Con.

Q.

R.

R.—*Con.*

S.

V.

W.

X.

Z.

LIST OF RECENT REPORTS OF GEOLOGICAL SURVEY.

Since 1910, reports issued by the Geological Survey have been called memoirs and have been numbered Memoir 1, Memoir 2, etc. Owing to delays incidental to the publishing of reports and their accompanying maps, not all of the reports have been called memoirs, and the memoirs have not been issued in the order of their assigned numbers, and, therefore, the following, list has been prepared to prevent any misconceptions arising on this account. The titles of all other important publications of the Geological Survey are incorporated in this list.

Memoirs and Reports Published During 1910.

REPORTS.

Report on a geological reconnaissance of the region traversed by the National Transcontinental railway between Lake Nipigon and Clay lake, Ont.—by W. H. Collins. No. 1059.

Report on the geological position and characteristics of the oil-shale deposits of Canada—by R. W. Ells. No. 1107.

A reconnaissance across the Mackenzie mountains on the Pelly, Ross, and Gravel rivers, Yukon and North West Territories—by Joseph Keele. No. 1097. Summary Report for the calendar year, 1909. No. 1120.

MEMOIRS—GEOLOGICAL SERIES.

MEMOIR 1. *No. 1, Geological Series.* Geology of the Nipigon basin, Ontario—by Alfred W. G. Wilson.

MEMOIR 2. *No. 2, Geological Series.* Geology and ore deposits of Hedley Mining district, British Columbia—by Charles Camsell.

MEMOIR 3. *No. 3, Geological Series.* Palæoniscid fishes from the Albert shales of New Brunswick—by Lawrence M. Lambe.

MEMOIR 5. *No. 4, Geological Series.* Preliminary memoir on the Lewes and Nordenskiöld Rivers coal district, Yukon Territory—by D. D. Cairnes.

MEMOIR 6. *No. 5, Geological Series.* Geology of the Haliburton and Bancroft areas, Province of Ontario—by Frank D. Adams and Alfred E. Barlow.

MEMOIR 7. *No. 6, Geological Series.* Geology of St. Bruno mountain, Province of Quebec—by John A. Dresser.

MEMOIRS—TOPOGRAPHICAL SERIES.

MEMOIR 11. *No. 1, Topographical Series.* Triangulation and spirit levelling of Vancouver island, B.C. 1909—by R. H. Chapman.

Memoirs and Reports Published During 1911.

REPORTS.

Report on a traverse through the southern part of the North West Territories from Lac Seul to Cat lake, in 1902—by Alfred W. G. Wilson. No. 1006.

Report on a part of the North West Territories drained by the Winisk and Upper Attawapiskat rivers—by W. McInnes. No. 1080.

Report on the geology of an area adjoining the east side of Lake Timiskaming—by Morley E. Wilson. No. 1004.

Summary Report for the calendar year 1910. No. 1170.

MEMOIRS—GEOLOGICAL SERIES.

MEMOIR 4. *No. 7, Geological Series.* Geological reconnaissance along the line of the National Transcontinental railway in western Quebec —by W. J. Wilson.

MEMOIR 8. *No. 8, Geological Series.* The Edmonton coal field, Alberta—by D. B. Dowling.

MEMOIR 9. *No. 9, Geological Series.* Bighorn coal basin, Alberta—by G. S Malloch.

MEMOIR 10. *No. 10, Geological Series.* An instrumental survey of the shorelines of the extinct lakes Algonquin and Nipissing in southwestern Ontario—by J. W. Goldthwait.

MEMOIR 12. *No. 11, Geological Series.* Insects from the Tertiary lake deposits of the southern interior of British Columbia, collected by Mr. Lawrence M. Lambe, in 1906—by Anton Handlirsch.

MEMOIR 15. *No. 12, Geological Series.* On a Trenton Echinoderm fauna at Kirkfield, Ontario—by Frank Springer.

MEMOIR 16. *No. 13, Geological Series.* The clay and shale deposits of Nova Scotia and portions of New Brunswick—by Heinrich Ries, assisted by Joseph Keele.

MEMOIRS—BIOLOGICAL SERIES.

MEMOIR 14. *No. 1, Biological Series.* New species of shells collected by Mr. John Macoun at Barkley sound, Vancouver island, British Columbia—by William H. Dall and Paul Bartsch.

Memoirs and Reports Published During 1912.

REPORTS.

Summary Report for the calendar year 1911. No. 1218.

MEMOIRS—GEOLOGICAL SERIES.

MEMOIR 13. *No. 14, Geological Series.* Southern Vancouver island—by Charles H. Clapp.

MEMOIR 21. *No. 15, Geological Series.* The geology and ore deposits of Phoenix, Boundary district, British Columbia—by O. E. LeRoy.

MEMOIR 24. *No. 16, Geological Series.* Preliminary report on the clay and shale deposits of the western provinces—by Heinrich Ries and Joseph Keele.

MEMOIR 27. *No. 17, Geological Series.* Report of the Commission appointed to investigate Turtle mountain, Frank, Alberta, 1911.

MEMOIR 28. *No. 18, Geological Series.* The geology of Steeprock lake, Ontario —by Andrew C. Lawson. Notes on fossils from limestone of Steeprock lake, Ontario—by Charles D. Walcott.

Memoirs and Reports Published During 1913.

REPORTS, ETC.

Museum Bulletin No. 1. Contains articles Nos. 1 to 12 of the Geological Series of Museum Bulletins, articles Nos. 1 to 3 of the Biological Series of Museum Bulletins, and article No. 1 of the Anthropological Series of Museum Bulletins.

Guide Book No. 1. Excursions in eastern Quebec and the Maritime Provinces, parts 1 and 2.

Guide Book No. 2. Excursions in the eastern Townships of Quebec and the eastern part of Ontario.

Guide Book No. 3. Excursions in the neighbourhood of Montreal and Ottawa.

Guide Book No. 4. Excursions in southwestern Ontario.

Guide Book No. 5. Excursions in the western peninsula of Ontario and Manitoulin island.

Guide Book No. 8. Toronto to Victoria and return via Canadian Pacific and Canadian Northern railways, parts 1, 2, and 3.

Guide Book No. 9. Toronto to Victoria and return via Canadian Pacific, Grand Trunk Pacific, and National Transcontinental railways.

Guide Book No. 10. Excursions in northern British Columbia and Yukon Territory and along the North Pacific coast.

MEMOIRS—GEOLOGICAL SERIES.

MEMOIR 17. *No. 28, Geological Series*. Geology and economic resources of the Larder Lake district, Ont., and adjoining portions of Pontiac county, Que.—by Morley E. Wilson.

MEMOIR 18. *No. 19, Geological Series*. Bathurst district, New Brunswick—by G. A. Young.

MEMOIR 26. *No. 34, Geological Series*. Geology and mineral deposits of the Tulameen district, B.C.—by C. Camsell.

MEMOIR 29. *No. 32, Geological Series*. Oil and gas prospects of the northwest provinces of Canada—by W. Malcolm.

MEMOIR 31. *No. 20, Geological Series*. Wheaton district, Yukon territory—by D. D. Cairnes.

MEMOIR 33. *No. 30, Geological Series*. The geology of Gowganda Mining division—by W. H. Collins.

MEMOIR 35. *No. 29, Geological Series*. Reconnaissance along the National Transcontinental railway in southern Quebec—by John A. Dresser.

MEMOIR 37. *No. 22, Geological Series*. Portions of Atlin district, B.C.—by D. D. Cairnes.

MEMOIR 38. *No. 31, Geological Series*. Geology of the North American Cordillera at the forty-ninth parallel, Parts I and II—by Reginald Aldworth Daly.

Memoirs and Reports Published During 1914.

REPORTS, ETC.

Summary Report for the calendar year 1912. No. 1305.

Museum Bulletin No. 2. Contains articles Nos. 13 to 18 of The Geological Series of Museum Bulletins, and article No. 2 of the Anthropological Series of Museum Bulletins.

Prospector's Handbook No. 1. Notes on radium-bearing minerals—by Wyatt Malcolm.

MUSEUM GUIDE BOOKS.

The Archæological collection from the southern interior of British Columbia—by Harlan I. Smith. No. 1290.

MEMOIRS—GEOLOGICAL SERIES.

MEMOIR 23. *No. 23, Geological Series.* Geology of the coast and islands between the Strait of Georgia and Queen Charlotte sound, B.C.—by J. Austen Bancroft.

MEMOIR 25. *No. 21, Geological Series.* Report on the clay and shale deposits of the western provinces (Part II) — by Heinrich Ries and Joseph Keele.

MEMOIR 30. *No. 40, Geological Series.* The basins of Nelson and Churchill rivers—by William McInnes.

MEMOIR 20. *No. 41, Geological Series.* Gold fields of Nova Scotia—by W. Malcolm.

MEMOIR 36. *No. 33, Geological Series.* Geology of the Victoria and Saanich map-areas, Vancouver island, B.C.—by C. H. Clapp.

MEMOIR 52. *No. 42, Geological Series.* Geological notes to accompany map of Sheep River gas and oil field, Alberta—by D. B. Dowling.

MEMOIR 43. *No. 36, Geological Series.* St. Hilaire (Beloeil) and Rougemont mountains, Quebec—by J. J. O'Neil.

MEMOIR 44. *No. 37, Geological Series.* Clay and shale deposits of New Brunswick—by J. Keele.

MEMOIR 22. *No. 27, Geological Series.* Preliminary report on the serpentines and associated rocks, in southern Quebec—by J. A. Dresser.

MEMOIR 32. *No. 25, Geological Series.* Portions of Portland Canal and Skeena Mining divisions, Skeena district, B.C.—by R. G. McConnell.

MEMOIR 47. *No. 39, Geological Series.* Clay and shale deposits of the western provinces. Part III—by Heinrich Ries.

MEMOIRS—ANTHROPOLOGICAL SERIES.

Memoirs in Press, August 15, 1914.